Collins

Instant
Revision

GCSE Science

Chris Sunley
Mike Smith
Series Editor: Jayne de Courcy

Contents

Published by HarperCollins*Publishers* Ltd
77–85 Fulham Palace Road
London W6 8JB

www.**Collins**Education.com
On-line support for schools and colleges

© HarperCollins*Publishers* Ltd 2001

First published 2001

ISBN 0 00 710973 3

Chris Sunley and Mike Smith assert the moral right to be identified as the authors of this work.

British Library Cataloguing in Publication Data
A catalogue record for this publication is available from the British Library

Edited by Eva Fairnell
Production by Kathryn Botterill
Cover design by Susi Martin-Taylor
Design by Gecko Limited
Printed and bound by Scotprint

Acknowledgements

Illustrations
Gecko Ltd

Every effort has been made to contact the holders of copyright material, but if any have been inadvertently overlooked, the Publishers will be pleased to make the necessary arrangements at the first opportunity.

You might also like to visit: www.**fire**and**water**.com
The book lover's website

Get the most out of your Instant Revision pocket book

1 **Maximise your revision time.** You can carry this book around with you anywhere. This means you can spend any spare moments dipping into it.

2 **Learn and remember what you need to know.** This book contains all the really important things you need to know for your exam. All the information is set out clearly and concisely, making it easy for you to revise.

3 **Find out what you don't know.** The *Check yourself* questions and *Score chart* help you see quickly and easily the topics you're good at and those you're not so good at.

What's in this book

1 The facts – just what you need to know

● There are sections covering all the Biology, Physics and Chemistry topics that you'll meet in your GCSE Science exam.

● The information is laid out in short blocks so that it is easy to read and remember.

● Tables and diagrams make facts easy to revise.

2 *Check yourself* questions – find out how much you know and boost your grade

● Each *Check yourself* is linked to one or more facts page. The numbers after the topic heading in the *Check yourself* tell you which facts page the *Check yourself* is linked to.

- The questions are graded in difficulty. They aren't actual exam questions but they will show you what you do and don't know.

- The reverse side of each *Check yourself* page gives you the answers **plus** tutorial help and guidance to boost your exam grade.

- There are points for each question. The total number of points for each *Check yourself* is always 20. When you check your answers, fill in the score box alongside each answer with the number of points you feel you scored.

3 The *Score chart* – an instant picture of your strengths and weaknesses

- *Score chart (1)* lists all the *Check yourself* pages.

- As you complete each *Check yourself*, record your points on the *Score chart*. This will show you instantly which areas you need to spend more time on.

- *Score chart (2)* is a graph which lets you plot your points against GCSE grades. This will give you a rough idea of how you are doing in each area. Of course, this is only a rough idea because the questions aren't real exam questions!

Use this Instant Revision pocket book on your own – or revise with a friend or relative. See who can get the highest score!

LIFE PROCESSES AND CELLS (1)

All living things (**organisms**), whether they are animals or plants, have the following characteristics:

- **Movement**: e.g. a dog running or a flower opening
- **Respiration**: releasing energy from food (this is NOT the same as breathing)
- **Sensitivity**: sensing and responding, e.g. a plant growing towards the light
- **Growth**: e.g. repairing a wound or a baby growing larger
- **Reproduction**: sexual (involving sex cells) or asexual (one parent and no sex cells)
- **Excretion**: getting rid of substances a body has made but does not need, e.g. humans breathing out carbon dioxide
- **Nutrition**: the need for food, e.g. plants make their food by photosynthesis and animals get theirs by eating.

Cells

All living things are made of cells.

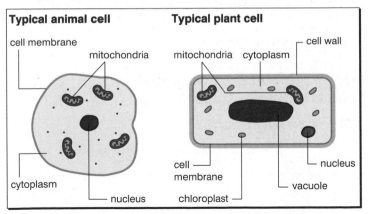

Typical animal cell

cell membrane
mitochondria
cytoplasm
nucleus

Typical plant cell

mitochondria   cytoplasm   cell wall
cell membrane
chloroplast
nucleus
vacuole

Cells are organised into **tissues**, which work together in **organs** which work together in organ **systems**.

Part of cell	Description
Nucleus	Contains chromosomes which carry genes (see *Genetics and Evolution* for more details) that control how cells grow and work
Cell membrane	Holds the cell together and controls substances entering and leaving the cell
Cytoplasm	Literally 'cell stuff', many chemical processes happen here
Mitochondria	Respiration happens in these
Cell wall	Made of cellulose and gives plant cells more rigid support than a membrane alone
Chloroplasts	Contain chlorophyll which absorbs the light energy needed for photosynthesis (see *Plants* for more details)
Vacuole	Large vacuole in plant cells contains cell sap; when it is full of liquid it helps to support the cell (animal cells may also contain small vacuoles)

Many cells are specialised to carry out particular jobs, e.g. sperm cells are specialised to swim to egg cells and fertilise them.

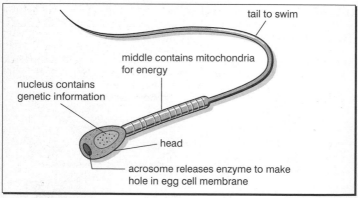

tail to swim

middle contains mitochondria for energy

nucleus contains genetic information

head

acrosome releases enzyme to make hole in egg cell membrane

Substances move in and out of cells in three main ways.

Diffusion

Particles in liquids and gases are constantly moving around. If a substance is more concentrated in one area than another (i.e. there is a concentration gradient) then the

The sugar molecules are concentrated in one area

The sugar molecules are spreading out

The sugar molecules are evenly concentrated

water molecule

random movement of the particles will cause an overall movement of the substance from the area of high concentration to that of low concentration, carrying on until the substance is evenly concentrated.

Osmosis

This is a special kind of diffusion when only the water in a solution passes through a partially permeable membrane from a dilute solution to more concentrated one.

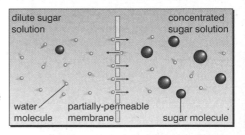

dilute sugar solution

concentrated sugar solution

water molecule

partially-permeable membrane

sugar molecule

Active Transport

Unlike diffusion and osmosis, active transport uses energy to transport particles across a cell membrane. Carrier molecules in the membrane move particles from one side

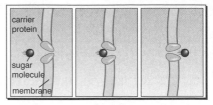

carrier protein

sugar molecule

membrane

to the other. Active transport can either move substances against a concentration gradient (from low to high concentrations) or speed up the movement that would happen by diffusion alone.

Respiration

The energy that living organisms need comes from the breakdown of food, usually glucose, by a process called respiration. This happens inside cells in the mitochondria. Respiration can be confused with breathing because oxygen is often involved, but they are NOT the same.

- **Aerobic respiration** is the release of energy from glucose using oxygen:

  glucose + oxygen → carbon dioxide + water + energy

  $$C_6H_{12}O_6 + 6O_2 \rightarrow 6CO_2 + 6H_2O + energy$$

- **Anaerobic respiration** is the release of energy from glucose without the use of oxygen. For example, this happens in muscles during exercise when oxygen cannot be supplied quickly enough for aerobic respiration.

  glucose → lactic acid + energy

Anaerobic respiration does not release as much energy from the same amount of glucose as aerobic respiration does.

The build up of lactic acid causes fatigue and muscle ache and eventually it has to be broken down using oxygen. This is why you still carry on breathing deeply *after* exercise to take in the necessary oxygen. This is called 'repaying the oxygen debt'.

Respiration causes changes to the air we breathe.

	Inhaled air	Exhaled air
Oxygen	21%	16%
Carbon dioxide	0.03%	4%
Nitrogen and other gases	79%	79%
Water	Variable	High
Temperature	Variable	High

Check yourself

Life Processes and Cells (1–4)

1 Name two cell components found in both animal and plant cells. **(1)**

2 Name two cell components only found in plant cells. **(1)**

3 Where in cells does respiration occur? **(1)**

4 Where is genetic information stored in a cell? **(1)**

5 Are the following examples of diffusion, osmosis or active transport:
(a) air freshener smell spreading around a room?
(b) a snail being dehydrated by being covered in salt? **(2)**

6 Which of diffusion, osmosis or active transport requires energy? **(1)**

7 What substance is energy normally released from in respiration? **(1)**

8 What type of respiration does not require oxygen? **(1)**

9 What waste product of respiration causes muscle tiredness? **(1)**

10 Describe how dissolved sugar particles will eventually spread out through a cup of tea, by diffusion, even if it is not stirred. **(2)**

11 How is osmosis different to other types of diffusion? **(2)**

12 What are the advantages of aerobic respiration over anaerobic respiration? **(2)**

13 What are the differences between breathing and respiration? **(2)**

14 The diagram shows an experiment that was set up to investigate respiration in mice. A pupil expected the water droplet to move to the left as the mouse used up the oxygen in the flask. In fact the droplet did not move at all. Explain why. **(2)**

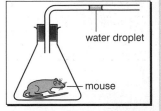

water droplet

mouse

1 Two from: nucleus, membrane, cytoplasm, mitochondria. (1)
   There are some exceptions, e.g. red blood cells contain no
   nucleus. Small vacuoles may also be found in animal cells.

2 Two from: cell wall, chloroplasts, large vacuole. (1)
   Again there are some exceptions, e.g. chloroplasts are only
   found in plant parts that are green and not in root cells.

3 In the mitochondria. (1)

4 In the nucleus. (1) 'Chromosomes' and 'DNA' would also be
   acceptable answers. Some questions may require these more
   specific answers.

5 (a) Diffusion. (1)   (b) Osmosis. (1) Substances moving
   from areas of high to low concentration show diffusion;
   water passing through a membrane shows osmosis.

6 Active transport. (1)
   The others are examples of passive transport.

7 Glucose. (1)
   If starved of glucose the body can use other foods instead.

8 Anaerobic. (1)

9 Lactic acid. (1)

10 Two from: sugar particles are continually moving (1);
   randomly (1); (so overall movement is) from areas of high
   concentration to areas of low concentration (1).
   These points apply to any example.

11 (Movement of) water (1); through a partially permeable
   membrane (1). Osmosis is still an example of diffusion.

12 More energy (for same amount of glucose) (1); lactic acid not
   produced / no oxygen debt (1). Anaerobic respiration will
   happen when there is not enough oxygen.

13 Breathing is moving air in and out of our bodies (1);
   respiration is releasing energy from food (1).
   Exam candidates often confuse these.

14 Aerobic respiration was happening (1); as much carbon
   dioxide was produced as oxygen used up (1).
   The balanced symbol equation for aerobic respiration shows
   equal numbers of carbon dioxide and oxygen molecules.

TOTAL

The Blood System

● What is blood made of?

Part of blood	What it looks like	Job
Plasma	Yellow liquid	Transports dissolved food, water, carbon dioxide, hormones, antibodies, urea, heat around body
Red blood cells		Transport oxygen around body using haemoglobin (as oxyhaemoglobin)
White blood cells		Fight disease by destroying invading microbes (see *Body Maintenance* for more information)
Platelets		Blood clotting

● What carries blood around the body?

Blood vessel	Job	Structure
Arteries	Carry blood at high pressure <u>a</u>way from the heart around the body	Thick muscular and elastic walls
Veins	Carry blood at low pressure from the body <u>in</u>to the heart	Thinner walls, valves
Capillaries	Exchange materials between the blood and body tissues	Very small with very thin, permeable walls

● Blood is pumped by the **heart**, which is made of four chambers: two **atria** to receive blood and two **ventricles** to pump the blood.

● The human blood system is a **double circulatory system** that allows the heart to maintain high blood pressure, ensuring efficient transport.

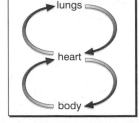

The Breathing System

Two processes occur:
- **ventilation** (breathing): moving air in and out of lungs
- **gaseous exchange**: taking oxygen into, and removing carbon dioxide from, the blood.

To breathe in (inhale):
- the diaphragm contracts and flattens
- the external intercostal muscles contract (and the internal intercostal muscles relax) making the ribs move forwards and upwards
- the volume of the thorax (chest) increases
- the air pressure inside the thorax drops
- greater air pressure outside the lungs causes air to enter.

To breathe out (exhale) the opposite happens.

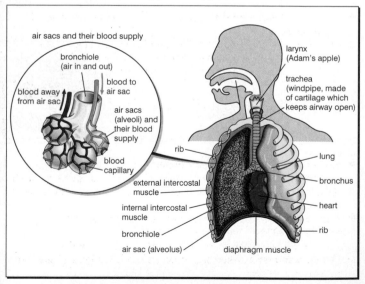

air sacs and their blood supply

bronchiole (air in and out)

blood to air sac

blood away from air sac

air sacs (alveoli) and their blood supply

blood capillary

larynx (Adam's apple)

trachea (windpipe, made of cartilage which keeps airway open)

rib

lung

bronchus

heart

rib

external intercostal muscle

internal intercostal muscle

bronchiole

air sac (alveolus)

diaphragm muscle

Food

Food type	Main functions	Some sources
Carbohydrate (starch and sugar)	Energy (respiration)	Bread, pasta, potatoes
Protein	Growth and repair	Meat, fish, eggs, beans
Fat	Energy store, insulation	Red meat, dairy products
Minerals and vitamins	Each has its own job, e.g. calcium for bones and teeth, vitamin C for healthy tissues	Calcium in milk, vitamin C in fruit
Fibre	Helps movement of food	Bran, vegetables

Digestion

As food passes through our digestive system it is **digested** so that it can pass out of the intestines into the blood and be carried to all parts of the body.

- **Mechanical** digestion is the breakdown of food into smaller pieces, e.g. by chewing.
- **Chemical** digestion is the breakdown of large food molecules (which are insoluble) into smaller ones (which are soluble). This is carried out by enzymes. An example is shown below.

Enzymes:

- are **specific**, each enzyme only breaks down one substance, e.g. protease digests protein, lipase digests fat, amylase digests starch, maltase, sucrase and lactase digest the sugars maltose, sucrose and lactose
- work best at a particular (**optimum**) temperature (usually around 35–40°C)
- work best at a particular (optimum) pH (usually around 7)
- can be **denatured** (irreversibly damaged) by too high a temperature or extremes of pH.

The Digestive System

What happens to food?

- **Ingestion** – food is taken into the mouth.
- **Digestion** – food is broken down into small molecules to make it soluble.
- **Absorption** – digested food is absorbed into the blood system by villi.
- **Assimilation** – food is taken in and used by the body's cells.
- **Egestion** – food that cannot be digested or absorbed leaves the body.

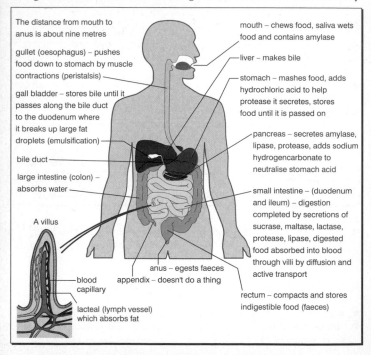

The distance from mouth to anus is about nine metres

gullet (oesophagus) – pushes food down to stomach by muscle contractions (peristalsis)

gall bladder – stores bile until it passes along the bile duct to the duodenum where it breaks up large fat droplets (emulsification)

bile duct

large intestine (colon) – absorbs water

A villus

blood capillary

lacteal (lymph vessel) which absorbs fat

appendix – doesn't do a thing

anus – egests faeces

mouth – chews food, saliva wets food and contains amylase

liver – makes bile

stomach – mashes food, adds hydrochloric acid to help protease it secretes, stores food until it is passed on

pancreas – secretes amylase, lipase, protease, adds sodium hydrogencarbonate to neutralise stomach acid

small intestine – (duodenum and ileum) – digestion completed by secretions of sucrase, maltase, lactase, protease, lipase, digested food absorbed into blood through villi by diffusion and active transport

rectum – compacts and stores indigestible food (faeces)

Human Body Systems (1–4)

1 Name two substances carried in blood plasma. (1)

2 (a) Which type of blood vessel contains valves? (1)

  (b) What is the job of valves? (1)

3 The aorta is one of the blood vessels carrying blood away from the heart. Is it an artery or a vein? (1)

4 Put the following in the order inhaled air passes through them: alveoli, bronchi, bronchioles, trachea. (1)

5 How does your diaphragm move when you inhale? (1)

6 Which food type provides most of our energy? (1)

7 Which of the following is an enzyme: bile, protease, hydrochloric acid? (1)

8 What is the job of the large intestine (colon)? (1)

9 What is egestion? (1)

10 Why are red blood cells a biconcave shape? (2)

11 Why does the left ventricle in the heart have a thicker wall than the right ventricle? (2)

12 Describe two ways alveoli are adapted for efficient transfer of oxygen into the blood. (2)

13 What is chemical digestion and why is it necessary? (2)

14 What are villi and what do they do? (2)

1 Two from: water, carbon dioxide, hormones, antibodies, urea, food/examples of food, e.g. amino acids, sugar, fat, protein, mineral ions, e.g. sodium, chloride. (1)
   Oxygen is carried by the red blood cells not the plasma.

2 (a) Veins. (1)
   (b) To stop the backflow of blood/make sure blood flows in the right direction. (1)
   Backflow might happen because the blood is at a low pressure.

3 Artery. (1)
   All arteries carry blood away from the heart.

4 Trachea, bronchi, bronchioles, alveoli. (1)

5 Downward. (1)
   It does this by contracting.

6 Carbohydrates. (1) We can also get some energy from other foods like fat and, in cases of starvation, from protein.

7 Protease. (1) The others are involved in digestion but they are not enzymes.

8 Absorbing water from material remaining from food. (1)

9 Removing indigestible material (faeces) from the body. (1) This is not the same as excretion (see *Body Maintenance*).

10 To provide a large surface area (1); to allow oxygen to enter/leave more easily (1). Having lots of small red blood cells rather than a few larger ones also gives a bigger surface area.

11 To pump blood harder/to provide a higher pressure (1); to pump blood all around the body (1).
   The right ventricle is only pumping blood around the lungs, which is a shorter trip requiring less pressure.

12 Two from: large surface area, thin, permeable, moist, good blood supply, concentration gradient (2).

13 Breakdown of large molecules into smaller ones (1); so the food is soluble/can be absorbed into the blood (1).
   In an exam do not say simply 'the breakdown of food' as this could apply to mechanical digestion, e.g. chewing.

14 Microscopic projections on the lining of the small intestine (1); absorb digested food into the blood (and lymph) (1).
   They have the same features as alveoli to aid absorption.

TOTAL

The Nervous System

Part		Job
Central nervous system (CNS)	Brain Spinal cord	Receive, process and send out information
Peripheral nerves	Sensory nerves	Carry signals to CNS
	Motor nerves	Carry signals from CNS
	Relay neurones	Connect other neurones

● Nerves are made of cells called **neurones**. Signals pass along them as **electrical impulses**. Signals pass across the gaps between neurones (**synapses**) carried by chemical **transmitter substances**.

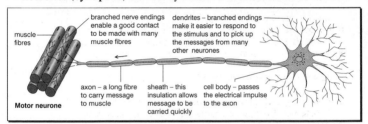

branched nerve endings enable a good contact to be made with many muscle fibres

dendrites – branched endings make it easier to respond to the stimulus and to pick up the messages from many other neurones

muscle fibres

axon – a long fibre to carry message to muscle

sheath – this insulation allows message to be carried quickly

cell body – passes the electrical impulse to the axon

Motor neurone

● **Sense organs** contain **receptor** cells that collect information. Each is sensitive to different types of **stimulus**. The **eye** contains receptors that detect light.

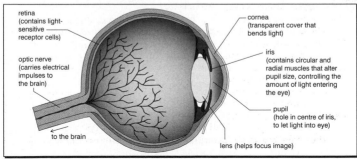

retina (contains light-sensitive receptor cells)

optic nerve (carries electrical impulses to the brain)

to the brain

cornea (transparent cover that bends light)

iris (contains circular and radial muscles that alter pupil size, controlling the amount of light entering the eye)

pupil (hole in centre of iris, to let light into eye)

lens (helps focus image)

● The nervous system controls how we respond to changes.
Automatic responses are called **reflexes**. During a reflex,
information flows along a pathway called a **reflex arc**.

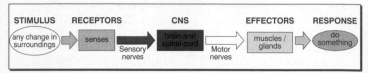

STIMULUS	RECEPTORS		CNS		EFFECTORS	RESPONSE
any change in surroundings	senses	Sensory nerves	brain and spinal cord	Motor nerves	muscles / glands	do something

The Endocrine (Hormone) System

Hormones are secreted by **endocrine** glands into the blood, which
carries them to their **target organs** where they have their effects.
Examples of hormones include the following:

● **Adrenaline**: produced by adrenal
glands, it prepares the body for 'flight
or fight' (increases heart rate,
breathing, sweating, release of
glucose into blood).

● **Growth hormone**: produced by the
pituitary gland, it encourages mental
and physical growth.

● **Insulin**: produced by the pancreas, it
travels to the liver where it promotes
the conversion of excess glucose to
glycogen, which is stored.

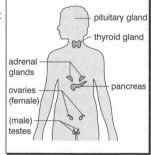

pituitary gland

thyroid gland

adrenal glands

ovaries (female)

(male) testes

pancreas

● **Testosterone**: produced by the testes, it controls **secondary sexual
characteristics** in males (hair grows on face and body, sperm
produced, voice breaks, growth spurt, sexual organs develop, body
becomes more muscular).

● **Oestrogen** and **progesterone**: produced by the ovaries, they
control secondary sexual characteristics in females (growth spurt,
breasts and sexual organs develop, menstrual cycle starts, hips
widen, pubic hair and hair under arms grows) and control changes
that occur during the menstrual cycle. They are used in fertility
drugs and the contraceptive pill.

Homeostasis

Homeostasis is maintaining the body's internal conditions.

Temperature is controlled by the **hypothalamus** in the brain, which monitors the blood temperature and controls the body's responses to maintain the body temperature at 37 °C:
- blood capillaries in the skin widen (**vasodilation**) or narrow (**vasoconstriction**) to control how much warm blood flows next to the surface
- **sweat** takes heat from the body as it evaporates
- hairs stand on end to trap a layer of insulating air, or lie flat to lose heat
- shivering generates heat
- the layer of fat under the skin acts as insulation.

Water is:
- gained in food, drink, respiration
- lost in urine, sweat, breathing, faeces.

The hypothalamus monitors the water content of the blood and controls how much water is excreted by the kidneys. Most of the contents of the blood are filtered out in the kidneys but then return to the blood, except **urea** (waste), excess water and salt.

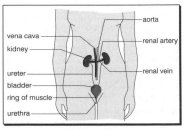

Carbon dioxide from respiration lowers the pH of the blood. This is detected by **chemoreceptors** in the blood vessels and the rate and depth of breathing is controlled to keep carbon dioxide levels in the blood low.

All examples of homeostasis involve **negative feedback** mechanisms that cause the body to respond to reduce the effect of any changes.

Body Defences

The body has defences to prevent the entrance of **pathogens** (disease-causing microbes):
- skin is a natural barrier
- **mucus** lining the airways traps air-borne pathogens, which are moved up to the top of the oesophagus by tiny hairs (**cilia**) on the epithelial (lining) cells
- acid in the stomach kills microbes taken in with food
- blood clots seal wounds, preventing infection.

The different types of **white blood cells** of the **immune system** attack pathogens that do enter the body:
- **phagocytes** surround, engulf and digest pathogens
- **lymphocytes** produce **antibodies** that destroy pathogens by binding to **antigens** on their surface.

Each type of pathogen has a particular type of antigen needing a **specific** antibody to bind to it. White blood cells can 'remember' how to make particular antibodies, which can be quickly made again to prevent a particular disease developing, so the body is '**immune**' to that disease. Immunity can be given artificially by injecting harmless forms of the pathogen, allowing the body to 'learn' how to make the right kind of antibodies without the disease.

Drugs can be used to help the body's defences, but some, legal and illegal, can affect the body badly.
- **Tobacco** contains **nicotine** (addictive, increases blood pressure and risk of heart disease), **tar** (causes lung cancer and damages the cilia leading to lung infections and smokers' cough) and releases **carbon monoxide** (causes breathlessness).
- **Alcohol** slows reactions and impairs judgement in the short term. Long-term use can harm the liver, brain and heart.
- **Solvents** slow brain activity, causing dizziness, loss of co-ordination and unconsciousness. They are also **toxic**.
- Some drugs are **addictive** (if you stop taking the drug you get **withdrawal symptoms**). Increased use of drugs increases **tolerance** (you need increased amounts to get the same effect).

Body Maintenance (1–4)

1 What is the job of the retina? (1)

2 Which of the following are reflexes: sneezing, blinking, blowing your nose, taking off your coat? (1)

3 Which type of neurones carry information into the CNS? (1)

4 If you have a fright and your heart rate increases, which hormone has been involved? (1)

5 What is meant by 'secondary sexual characteristics'? (1)

6 If you were sweating a lot on a very hot day how would you expect your urine to be different to that on a cooler day? (1)

7 Where would you find the highest levels of urea, in the renal artery or the renal vein? (1)

8 What is an antigen? (1)

9 How do phagocytes attack pathogens? (1)

10 Which substance in tobacco smoke increases the risks of lung cancer? (1)

11 How does the pupil get larger in dim light and why is this important? (3)

12 After a meal how does insulin released from the pancreas restore the blood glucose levels? (2)

13 What happens in vasodilation? (3)

14 How is foreign material removed from the lungs? (2)

1 To sense light/ convert light energy into electrical impulses. (1)

2 Sneezing and blinking. (1)
   Reflexes are automatic and usually protective in some way.

3 Sensory. (1)

4 Adrenaline. (1)
   Adrenaline has many target organs, unlike some hormones.

5 Those body changes that occur at puberty. (1)
   In an exam you might also get marks for giving examples.

6 Smaller amount/more concentrated/darker/more yellow. (1)
   If more water than usual has been lost in sweating then there will be less in the urine.

7 Renal vein. (1)
   All veins take blood away from the body's organs towards the heart.

8 A chemical on the surface of microbes/pathogens. (1)
   Remember each type of pathogen has its own type of antigen. In fact every cell carries antigens but exam questions about them will be referring to pathogens.

9 Engulf them. (1)
   This is also called phagocytosis.

10 Tar. (1)

11 Radial muscles contract (1); circular muscles relax (1); this lets more/enough light into eye to see (1).
    The opposite happens in bright light.

12 Two from: (insulin) travels in blood to the liver; converts excess glucose to glycogen; glycogen stored in liver. (2)
    In some exam questions you might need all these points to get full marks.

13 Blood capillaries/vessels near surface widen (1); more blood flows near to the surface (1); increased energy transfer as heat (1).
    It is a very common mistake for exam candidates to describe the blood vessels moving towards the surface.

14 Trapped by mucus (1); cilia move mucus upwards (1).
    Eventually it passes down the oesophagus.

TOTAL

PLANTS (1)

Photosynthesis

Plants need food just like animals, but unlike animals they make it for themselves by photosynthesis:

$$\text{carbon dioxide} + \text{water} \xrightarrow[\text{light}]{\text{chlorophyll}} \text{glucose} + \text{oxygen}$$

$$6CO_2 \qquad\qquad 6H_2O \qquad\qquad\qquad C_6H_{12}O_6 \qquad 6O_2$$

- Glucose is used for respiration or converted into other useful substances, e.g. sucrose (stored in fruit), starch (stored in seeds, roots, shoots), cellulose (for cell walls), oils (stored in seeds) and proteins (for growth).
- Increasing the carbon dioxide concentration, light intensity and temperature can all increase the rate of photosynthesis up to the point when insufficient levels of one of the other factors needed acts as a **limiting factor**.
- **Leaves** are adapted to be efficient at photosynthesis.

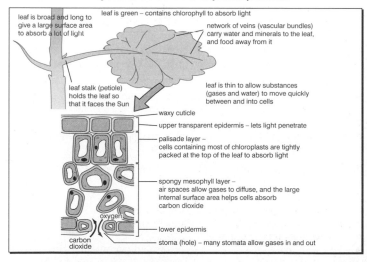

leaf is broad and long to give a large surface area to absorb a lot of light

leaf is green – contains chlorophyll to absorb light

network of veins (vascular bundles) carry water and minerals to the leaf, and food away from it

leaf stalk (petiole) holds the leaf so that it faces the Sun

leaf is thin to allow substances (gases and water) to move quickly between and into cells

waxy cuticle

upper transparent epidermis – lets light penetrate

palisade layer – cells containing most of chloroplasts are tightly packed at the top of the leaf to absorb light

spongy mesophyll layer – air spaces allow gases to diffuse, and the large internal surface area helps cells absorb carbon dioxide

lower epidermis

oxygen

carbon dioxide

stoma (hole) – many stomata allow gases in and out

Transport

Plants transport materials through:
● **xylem** vessels, which carry water and dissolved minerals from the roots to the leaves (xylem also helps support the plant)
● **phloem** vessels, which carry dissolved food (by **translocation**), mainly sucrose, from the leaves to growing and storage regions.

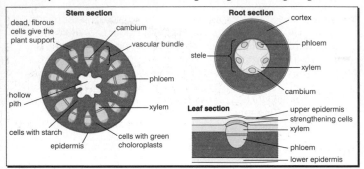

● Water enters the **root hair cells** by **osmosis** and moves into the xylem vessels where the water is pulled up in the **transpiration stream**.

Transpiration:
● is the loss of water from the leaves as it evaporates into the air spaces inside and then diffuses out through the stomata
● happens most quickly when it is warm, windy, dry and light
● cools plants and brings up minerals from the soil, but too much can cause a plant to dehydrate and **wilt**.

Plants have different ways of reducing too much **water loss**:
● a waxy cuticle on leaves
● hairs on leaves to reduce evaporation
● stomata located on the cooler underside of leaves
● leaves reduced to spines, exposing less surface area
● stomata sunken below the leaf surface
● closing stomata (guard cells lose water and become **flaccid**).

Turgor and Plasmolysis

When a plant wilts the cells lose water and become flaccid (less firm). If cells lose so much water that the cell membrane shrinks away from the cell wall, the cells are **plasmolysed**. Healthy cells full of water are **turgid** (firm), which means that they press against each other supporting the plant.

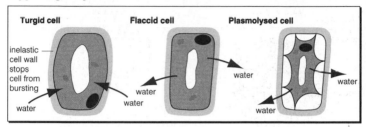

Minerals

To grow properly plants need certain minerals from the soil. The table shows some of these.

Mineral	Element	Use in plant	Problems caused by deficiency
Nitrates	Nitrogen	To make proteins to make new cells	Poor growth Pale/yellow leaves
Phosphates	Phosphorous	Involved in respiration	Poor growth Small leaves Low fruit yield
Potassium salts	Potassium	Control of salt balance in cells	Mottled leaves Low fruit yields
Magnesium salts	Magnesium	To make chlorophyll	Yellow patches on leaves

- Plants take in minerals from the soil by **active transport** because the minerals may only be present in the soil in small amounts and so may need to be absorbed against a concentration gradient.
- Farmers and gardeners may add **fertilisers** to replace minerals lost from the soil.

Plant Hormones

Plant hormones (or **plant growth regulators**) control many of the ways plants grow and develop, e.g. growth of roots, shoots and buds, flowering, fruit formation and ripening, germination, leaf fall, healing wounds.

Many plant hormones are used commercially, e.g. for:
● killing weeds in a lawn but not the grass (selective weedkillers)
● encouraging root growth in cuttings (rooting powder)
● storing potatoes and cereals for longer by delaying germination
● reducing damage when transporting soft fruit and vegetables by delaying ripening
● making crops ripen at the same time for ease of harvesting.

Tropisms

Tropisms are directional growth responses to certain stimuli, e.g. the growth of shoots towards light (**phototropism**) or the growth of roots downwards (**geotropism**). These tropisms are controlled by a plant hormone called **auxin**, which is made in the tips of shoots and roots and moves in solution to other parts.

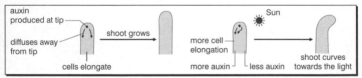

● If light shines on a shoot from one side auxin moves to the dark side, where it causes the cells to grow larger.
● Gravity causes auxin to collect on the lower side of roots, where it *stops* cells getting larger.

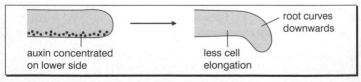

Plants (1–4)

1 Complete the word equation for photosynthesis:
carbon dioxide + ? → glucose + ? (1)

2 Where does carbon dioxide enter leaves? (1)

3 In which cells does most photosynthesis occur? (1)

4 Which vessels carry dissolved food through a plant? (1)

5 What is transpiration? (1)

6 How is the surface area of roots increased to increase absorption? (1)

7 Why do plants need nitrates? (1)

8 Why does lack of magnesium cause leaves to yellow? (1)

9 What is geotropism? (1)

10 Soft fruit can be treated with certain plant hormones to delay their ripening. Why would this be done? (1)

11 A tomato grower gives her crop extra carbon dioxide because this will increase the rate of photosynthesis. The results of this are shown in the graph.

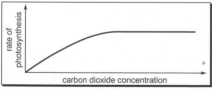

After a certain point adding more carbon dioxide will not increase the rate of photosynthesis any further. Why not? (2)

12 Explain why it is an advantage for a leaf to have its stomata on its lower side. (2)

13 Describe what happens to a plant's cells when it wilts. (3)

14 Seedlings growing on a window sill will grow towards the light outside. Describe what makes them grow this way. (3)

1 carbon dioxide + water → glucose + oxygen (1) **On a higher
   level paper be prepared to give the symbol equation.**
2 Through the stomata. (1) **Do not make the common mistake
   of thinking that water also enters through the stomata.**
3 The cells of the palisade layer. (1)
   **Some photosynthesis will also happen in the spongy layer
   and the guard cells, but they contain fewer chloroplasts.**
4 Phloem. (1)

5 Loss of water from leaves. (1)
   **The 'transpiration stream' is the flow of water from the roots,
   up the stem and out through the leaves.**
6 Root hair cells. (1)

7 Growth/ to make new cells/ to make proteins. (1)
   Always try to give as detailed an answer as possible.
8 Magnesium is needed to make chlorophyll. (1)

9 Plant growth in response to gravity. (1) **Roots are positively
   geotropic and shoots are negatively geotropic.**
10 The fruits would be less likely to be damaged in transit. (1)
   **Expect to get some questions about the practical
   applications of biological ideas.**
11 Carbon dioxide is being absorbed as quickly as possible (1);
   there is a limiting factor, e.g. light levels (1).
   Two marks are given here for two separate ideas.
12 It will reduce water loss (by evaporation/ transpiration) (1); as
   the lower side is cooler (1). **This is different to question 11
   in that the two marks are given for one idea that is expanded on.**
13 Cells lose water (1); cell contents shrink/ plasmolysis (1);
   cells become less firm/ flaccid (1).
   **On a higher level paper expect to have to explain wilting in
   terms of what is happening to the cells and not just with
   vague references to a lack of water.**
14 Auxin (1); accumulates on the dark side (of the shoot) (1);
   increased growth on dark side (1).
   **Three marks for a question tells you that a detailed answer is
   required. Use diagrams if it helps you to answer.**

TOTAL

Ecology is the study of the relationships between living things and their environment.

Competition

There is always a struggle for survival among living things because of competition for limited resources.

- Animals may not survive through lack of food or water, disease, accidents, weather, predators.
- Plants may not survive through lack of light, water or minerals, disease, weather, being eaten.

The struggle for survival prevents populations from constantly increasing.

Predation

One example showing how predation can affect population size is that of lynx and snowshoe hares in northern Canada.

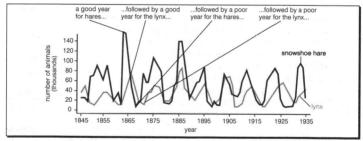

Human Population

The human population is increasing **exponentially** (the rate of increase is increasing) because of:

- increased food availability
- medical improvements, e.g. immunisations and antibiotics
- improved living conditions.

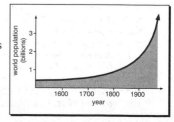

Food Chains

 Sun

 Grass: **producer**. Green plants make food in photosynthesis using energy from the Sun (see *Plants*).

 Grasshopper: **primary consumer** (or **herbivore**) (animals that eat plants or parts of plants).

 Vole: **secondary consumer** (or **carnivore** or **predator**) (animals that eat primary consumers).

 Owl: **tertiary consumer** (also carnivore or predator) (animals that eat secondary consumers).

- Animals that are hunted by predators are called **prey**.
- Some animals are **omnivores**, eating animals and plants.
- The stages of a food chain are also called **trophic levels**.
- Food chains are usually linked with others to form **food webs**.
- Pyramids are another way of describing food chains or webs.

Pyramids of number show the relative numbers of each type of living thing in an ecosystem. **Pyramids of biomass** show the mass of living material at each stage.

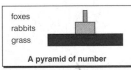

foxes
rabbits
grass

A pyramid of number

Only a small proportion of the energy at each trophic level is passed on to the next. Only energy that is used for growth is available to consumers. This is why food chains are usually no more than four or five stages long and pyramids of biomass always get smaller as you go up.

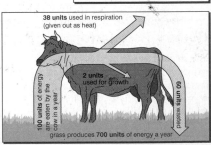

38 units used in respiration (given out as heat)

2 units used for growth

60 units wasted

100 units of energy are eaten by the cow in a year

grass produces **700 units** of energy a year

● There are only limited amounts of the chemical elements that living things need and are made of. The only way that living things can continue to get these elements is if they are recycled. Two such elements are carbon and nitrogen.

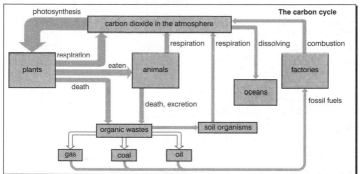

The carbon cycle

photosynthesis

carbon dioxide in the atmosphere

respiration — respiration — dissolving — combustion

respiration

plants — eaten → animals — factories

death

oceans

death, excretion — fossil fuels

organic wastes — soil organisms

gas — coal — oil

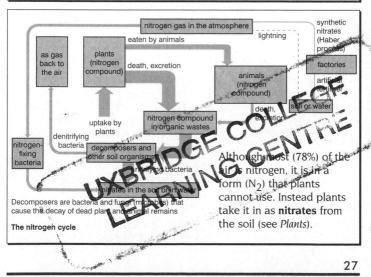

nitrogen gas in the atmosphere

synthetic nitrates (Haber process)

lightning

as gas back to the air

plants (nitrogen compound) — eaten by animals

death, excretion

animals (nitrogen compound)

factories

artificial fertiliser

death, excretion — soil or water

uptake by plants

nitrogen compound in organic wastes

nitrogen-fixing bacteria

denitrifying bacteria

decomposers and other soil organisms

nitrifying bacteria

nitrates in the soil or in water

Decomposers are bacteria and fungi (microbes) that cause the decay of dead plant and animal remains

The nitrogen cycle

Although most (78%) of the air is nitrogen, it is in a form (N_2) that plants cannot use. Instead plants take it in as **nitrates** from the soil (see *Plants*).

Human Influences on the Environment

An increasing population and technological advances mean that more resources (land, raw materials, food, energy sources) are being used and more waste (**pollution**) is being caused.

Agrochemicals are chemicals used by farmers to increase crop production:

- **Insecticides** (pesticides) kill insect pests. Some, such as DDT, are banned in many countries because they are **persistent** (do not break down) and can pass along food chains. Alternatives include using insecticides that do break down or using predator species (**biological control**).
- **Herbicides** kill weeds (unwanted plants) but these plants may be food for animals.
- **Artificial fertilisers** are chemicals added to improve crop growth. They can **leach** through the soil when it rains and get into streams, river and lakes, where they encourage the growth of algae (small green plants) on the surface which kills water plants by blocking light. As the dead plants decay the bacteria responsible use up oxygen from the water, killing fish and other animals. This is called **eutrophication**. It can be reduced by using natural fertilisers, like manure or compost, or crop rotation.

Natural **greenhouse gases** in the atmosphere allow energy from the Sun to warm the Earth but reduce heat loss back into space. Levels of greenhouse gases such as carbon dioxide, methane and CFCs are increasing because of pollution and may lead to an increased **greenhouse effect** which could cause **global warming**.

Sulphur dioxides and other gases released when fossil fuels are burnt, combine with water in the atmosphere making rain acidic. **Acid rain** harms plants and also washes useful minerals from the soil. Aluminium washed into rivers poisons fish.

The **ozone layer** in the atmosphere reduces the amount of harmful UV radiation reaching the Earth. Increased amounts of CFC gases can cause the ozone layer to break down (**ozone depletion**).

Ecology and the Environment (1–4)

1 Grass plants growing in a field are competing for several resources. Name two. (1)

2 (a) Complete the food chain by adding arrows:
grass   antelopes   lions. (1)
   (b) Which organism in the chain is the producer? (1)

3 (a) Which bacteria convert atmospheric nitrogen into nitrates? (1)
   (b) Why do plants need nitrates? (1)

4 Give two ways carbon dioxide is put into the atmosphere. (2)

5 What are pesticides? (1)

6 Describe one problem that could be caused by global warming. (1)

7 Why is the ozone layer important? (1)

8 (a) Draw a pyramid of numbers for the food chain:
oak tree → caterpillars → blue tits (1)
   (b) Draw a pyramid of biomass for the same food chain. (1)
   (c) Why doesn't all the energy from the caterpillars pass to the blue tits? Give two reasons. (2)

9 Describe how fertilisers leaching into rivers can cause the death of fish. (3)

10 Describe how carbon from the atmosphere becomes part of the body of an animal. (3)

1 Two from: light/ water/ minerals. (1) A common exam response would be 'space' but this is not a very good answer.

2 (a) grass → antelopes → lions (1)
Many exam candidates put arrows the wrong way round. The arrows show the way energy is passing.

   (b) Grass. (1) Plants produce their own food by photosynthesis. Animals get food by eating other things.

3 (a) Nitrogen-fixing bacteria. (1)
   (b) To make protein. (1) Protein is needed e.g. for growth.

4 Respiration. (1) Combustion. (1)

5 Chemicals that kill pests (organisms that damage the crops, usually by eating them). (1) Pesticides are usually insecticides but include others such as fungicides that kill fungal pests.

6 Climate or weather changes/ flooding. (1)

7 It reduces the amount of UV (ultraviolet) radiation reaching the Earth's surface. UV radiation causes tanning but also skin cancer. (1) Many exam candidates get confused between ozone damage and the greenhouse effect.

8 (a) blue tits, caterpillars, oak tree (1) The 'inverted' pyramid is caused by the oak tree being much larger than the other organisms involved.

   (b) blue tits (1) Pyramids of biomass are caterpillars, oak tree — always pyramid shaped.

   (c) Two from: not all the caterpillars are eaten/ other things eat the caterpillars/ some of the caterpillars' bodies is egested/ respiration by the caterpillars. (2) These points would apply to other examples as well.

9 Growth of algae (1); lack of light kills water plants (1); bacteria rotting the plants use up oxygen (1). A common error would be to simply say that the fertilisers kill the fish.

10 Plants take in CO_2 during photosynthesis (1); animals eat the plants (1); digested plant material is used for growth (1).

TOTAL

Living things, even of the same species, are different to each other (**variation**). Variation can be **continuous** (having any value in a range), e.g. height, or **discontinuous**, e.g. gender or blood group. It can be caused by the **environment** (e.g. diet, climate) or by **genes** (i.e. **inherited** from parents).

Genes:
- are the instructions that control our features and are found inside the cell **nucleus** carried by **chromosomes**
- are made of a chemical called **DNA** (deoxyribonucleic acid), along which is a series of four chemical bases called A, C, T and G for short, the sequence of bases spells out a code which is how genes carry information
- work by telling cells how to make particular proteins.

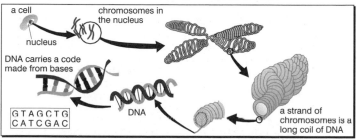

- When cells divide chromosomes make copies of themselves (**replicate**). There are two types of cell division.
- Sometimes chromosomes and genes are not copied exactly and the instructions become altered. These are **mutations** and may happen naturally or be caused e.g. by radiation.

Mitosis	Meiosis
Occurs in normal cell growth and **asexual** reproduction.	Only occurs when sex cells (e.g. eggs and sperm) are made.
New cells have the same chromosome number as the original and are genetically identical to it and each other.	New cells have half the chromosome number as the original. They are genetically different to it and each other.

Inheritance

Human body cells contain **23 pairs** of chromosomes (46 in total), except for egg and sperm cells which contain one of each chromosome pair (23 in total). At **fertilisation**, eggs and sperm join together and the full chromosome number is restored.

Except for the sex cells, cells contain two copies of each gene, one on each of a pair of chromosomes (called **homologous** chromosomes). For many genes there are different versions (**alleles**). For example, being able to roll your tongue is determined by a gene which has two alleles, called R (roller) and r (non-roller). R is **dominant** and r is **recessive** which means that if you have both of them you will be able to roll your tongue. We can show how alleles are inherited by using a genetic diagram:

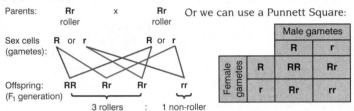

	Male gametes	
	R	**r**
R	RR	Rr
r	Rr	rr

Female gametes

This is a **monohybrid** cross because only one feature, tongue rolling ability, is being considered.

Some inheritance terms are:
- **genotype**: an individual's combination of alleles, e.g. Rr
- **phenotype**: an individual's features, e.g. roller
- **heterozygous**: having different alleles, e.g. Rr
- **homozygous**: having identical alleles, e.g. RR or rr.

Gender is decided by the **sex chromosomes**, X and Y. Human females have XX and males have XY.

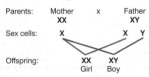

Understanding inheritance and genes has allowed people to alter the genes and features of other living things.

Selective Breeding (Artificial Selection)

One example is that modern varieties of cows produce more milk than their wild ancestors. This has been done by:

- selecting those cows producing more milk than the others
- using these cows for breeding
- selecting the offspring that produce the most milk
- repeating this for many generations.

Cloning

If you have a plant that has features you want, you can produce offspring that are genetically identical to the original by cloning them, e.g. by taking cuttings or by **tissue culture**:

- cut many small pieces from the plant you want
- sterilise the pieces in mild bleach to kill any microbes
- put the pieces on growth medium (nutrients and hormones)
- later use other growth media and eventually normal compost.

Scientists have also been able to clone some animals.

Genetic Engineering

This means taking the genes from one organism and inserting them into another. Human insulin, used to treat diabetes, is made by bacteria that contain the human insulin gene.

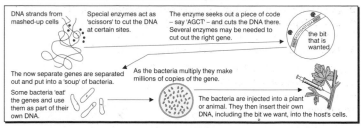

DNA strands from mashed-up cells

Special enzymes act as 'scissors' to cut the DNA at certain sites.

The enzyme seeks out a piece of code – say 'AGCT' – and cuts the DNA there. Several enzymes may be needed to cut out the right gene.

the bit that is wanted

The now separate genes are separated out and put into a 'soup' of bacteria.

As the bacteria multiply they make millions of copies of the gene.

Some bacteria 'eat' the genes and use them as part of their own DNA.

The bacteria are injected into a plant or animal. They then insert their own DNA, including the bit we want, into the host's cells.

Evolution

Over millions of years many animals and plants have evolved into new forms that have been better adapted for survival. Some species do not appear to have changed very much for millions of years because they were already very well adapted and their environments have changed little. Some species have become **extinct** because they were not able to survive or evolve in changing conditions.

Evidence for evolution is found from:
- **fossils**, although there are gaps in the fossil record (because fossilisation is a rare event and many fossils may lie undiscovered or have been destroyed e.g. by erosion)
- **affinities**, i.e. similarities between different species that are most easily explained if they are related (evolved from a common ancestor).

Natural Selection is a theory developed by Charles Darwin and Alfred Wallace that explains how evolution can happen:
- Living things produce more offspring than will ever survive into maturity. Competition for limited resources means that most will die before they have a chance to reproduce. Within a species there is always variation. Some individuals will have features that help them to survive better than others (**survival of the fittest**).
- Those individuals who have the best features are more likely to live longer and reproduce more. If the features that help survival are genetically controlled they may be passed onto offspring who in turn will be more likely to survive than others.
- If this process continues over very many generations the population will change (evolve) to become better adapted to survive.

Genetics and Evolution (1–4)

1 Which of the following are examples of continuous variation: nose length, gender, weight, blood group? (1)

2 What are genes made of? (1)

3 Where, in a cell, are genes found? (1)

4 Human skin cells contain 46 chromosomes. How many would there be in:

   (a) a muscle cell?

   (b) a sperm cell? (2)

5 What type of cell division is involved when you grow new skin to repair a cut? (1)

6 What are alleles? (1)

7 What is the difference between being homozygous and heterozygous for a particular gene? (1)

8 Which sex chromosomes are found in:

   (a) egg cells?

   (b) sperm cells? (2)

9 What are the chances of a baby being a boy or a girl? (1)

10 Cystic fibrosis is an inherited disease. The allele (gene) **f** for cystic fibrosis is recessive to the allele **F** for the healthy condition. Draw a genetic diagram to show how two parents who do not show the disease could have a child who does. (3)

11 Describe how you would use selective breeding to produce a breed of sheep that has a thicker coat of wool. (3)

12 Cactus plants have long roots to reach water in desert conditions. Describe the steps by which they could have evolved from ancestors with shorter roots. (3)

1 Nose length, weight. (1)
The others show discontinuous variation.

2 DNA. (1)

3 Nucleus/chromosome. (1)

4 **(a)** 46. (1) **(b)** 23. (1)
Eggs also have 23, so a fertilised egg will have 46.

5 Mitosis. (1) Virtually all cell division is mitosis. Only sex cells are formed by meiosis.

6 Different versions of a gene. (1)
Some genes may have several alleles or only one.

7 Homozygous means having identical alleles, e.g. RR; heterozygous means having two different alleles, e.g. Rr. (1)

8 **(a)** X only. (1) **(b)** X or Y. (1)

9 50%. (1) All eggs carry an X. Half of the sperm cells carry an X and half a Y.

10 Parents: **Ff** x **Ff** (3)
Sex cells: **F f** **F f**

Offspring: **FF Ff** **Ff ff**
Healthy Cystic fibrosis

The parents are **Ff** since they each pass on one **f** to the child with the disease, but they do not show the disease themselves.

11 Choose sheep with the thickest coats of wool (1); use these for breeding (1); choose offspring that have the thickest coats and repeat the process many times (1).
In exams the examples will differ but the steps are the same.

12 Cactus plants with slightly longer roots are more likely to survive (1); these will be more likely to reproduce and pass on the genes for longer roots to their offspring (1); if this is repeated over many generations the average length of cactus roots will increase (1). Again, although examples will differ, the steps are the same. Notice how the steps are also basically the same as selective breeding except it is not a human who is selecting which are most likely to breed.

TOTAL

Combining Powers

- When elements combine together chemically they form **compounds**.
- Before the formula of a compound can be worked out **the combining powers** of the elements in the compound need to be known. In most cases these can be worked out from the position of the element in the periodic table (see *The Periodic Table*).

Group number	1	2	3	4	5	6	7	0
Combining power	1	2	3	4	3	2	1	0

- The combining powers of the transition metals are usually given when the compound is named. For example, in the compound iron(III) chloride the iron has a combining power of 3, whereas in iron(II) chloride the iron has a combining power of 2.
- Where an element has a combining power different to that expected from its position in the periodic table, the combining power is given as above. For example, in sulphur(VI) oxide the sulphur has a combining power of 6.
- The combining power of hydrogen is 1.

Writing the Formulae of Simple Compounds

1 Write down the name of the compound.
2 Write down the chemical symbols for the elements.
3 Write the combining power of each element.
4 Cancel the numbers if possible.
5 Change over the combining powers. Write them after the symbol, slightly below the line.
6 There is no need to write '1's in the formula.

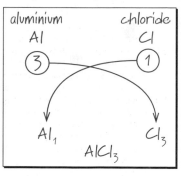

Writing the Formulae of more Complicated Compounds

● Some elements commonly exist together as a group called a **radical**. For example, in a carbonate radical there is 1 carbon atom and 3 oxygen atoms, CO_3. The radical cannot exist on its own, it must combine with another element or radical.

● The combining powers of the most common radicals are given below.

Combining power 1		Combining power 2		Combining power 3	
Hydroxide	OH	Carbonate	CO_3	Phosphate	PO_4
Nitrate	NO_3	Sulphate	SO_4		
Hydrogen carbonate HCO_3					

● The same rules apply as before:

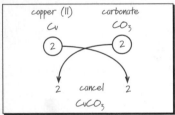

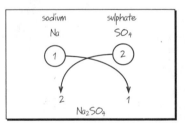

● If the formula contains more than one radical unit, the radical must be put in brackets. For example:

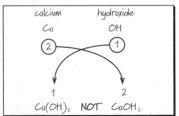

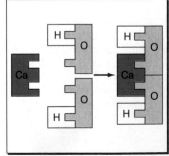

Writing Chemical Equations

- In a chemical reaction the starting chemicals are called the **reactants** and the finishing chemicals are known as the **products**.
- The rules for writing an equation are:

1 Write down the word equation.
2 Write down the symbols for the elements and formulae for the compounds.
3 Balance the equation (making sure that there are the same number of each type of atom on each side of the equation).

calcium carbonate	$\rightarrow$	calcium oxide	+	carbon dioxide
$CaCO_3$	$\rightarrow$	CaO	+	CO_2

In this example the equation already balances.

- When writing equations it is important to remember that many elements are **diatomic**, i.e. they exist as molecules containing two atoms.

1	hydrogen	+	oxygen	$\rightarrow$	water
2	H_2	+	O_2	$\rightarrow$	H_2O
3	$2H_2$	+	O_2	$\rightarrow$	$2H_2O$

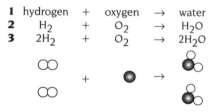

Element	Molecule
Hydrogen	H_2
Oxygen	O_2
Nitrogen	N_2
Chlorine	Cl_2

Ionic Equations

- Ionic equations, as the name suggests, are used for showing reactions between **ions** (charged atoms or molecules). The size of the charge on an ion is the same as the combining power. Metals and hydrogen form positive ions whereas non-metals usually form negative ions.
- In many ionic reactions some of the ions play no part – they are known as **spectator ions**. The spectator ions do not need to be included in the equation. For example:

copper(II) + sodium   → copper(II) + sodium
sulphate   hydroxide   hydroxide   sulphate

$$Cu^{2+}(aq) + 2OH^-(aq) \rightarrow Cu(OH)_2(s)$$

In this reaction the sodium and sulphate ions are spectator ions.

lead(II)   + sodium   → lead(II)   + sodium
nitrate   hydroxide   hydroxide   nitrate

$$Pb^{2+}(aq) + 2OH^-(aq) \rightarrow Pb(OH)_2(s)$$

The equation must balance both in terms of symbols and charges.

Half Equations

- Half equations are used to summarise the reactions that take place at the electrodes during electrolysis. They must balance in terms of symbols and charges. The symbol e^- is used to represent an electron.
- Positive ions are attracted to the cathode and gain electrons, e.g. copper ions are discharged as copper.
$$Cu^{2+}(aq) + 2e^- \rightarrow Cu(s)$$
e.g. aluminium ions are discharged as aluminium
$$Al^{3+}(aq) + 3e^- \rightarrow Al(s)$$
- Negative ions are attracted to the anode and lose electrons, e.g. oxide ions are discharged as oxygen.
$$2O^{2-}(l) \rightarrow O_2(g) + 4e^-$$

Formulae and Equations (1–4)

1 Calcium is an element in group 2. What is its combining power? (1)

2 Sulphur is an element in group 6. What is its combining power? (1)

3 Work out the chemical formulae of the following compounds:
   (a) magnesium oxide. (1)
   (b) hydrogen chloride. (1)
   (c) calcium bromide. (1)
   (d) copper(II) fluoride. (1)

4 Work out the chemical formulae of the following compounds:
   (a) nickel(II) sulphate. (1)
   (b) hydrogen nitrate (nitric acid). (1)
   (c) iron(II) hydroxide. (1)
   (d) aluminium carbonate. (1)

5 Balance the following equations:
   (a) $Ca + O_2 \rightarrow CaO$ (1)
   (b) $N_2 + H_2 \rightarrow NH_3$ (1)

6 Write balanced symbol equations from the following word equations:
   (a) magnesium + hydrogen chloride $\rightarrow$ magnesium chloride + hydrogen (1)
   (b) methane (CH_4) + oxygen $\rightarrow$ carbon dioxide + water (1)

7 Write an ionic equation to show the formation of lead(II) hydroxide from lead(II) ions and hydroxide ions. (2)

8 Write half equations for the following reactions:
   (a) The discharge of aluminium ions at the cathode. (2)
   (b) The discharge of chloride ions at the anode. (2)

1 2. (1)
The combining power is the same as the group number.

2 2. (1)
The combining power is 8 – the group number = 2.

3 (a) MgO. (1) Both Mg and O have combining powers of 2.
   (b) HCl. (1) Both H and Cl have combining powers of 1.
   (c) $CaBr_2$. (1) Ca has a combining power of 2; Br has a combining power of 1.
   (d) CuF_2. (1) Cu has a combining power of 2; fluorine has a combining power of 1.

4 (a) $NiSO_4$. (1) Nickel and the sulphate radical both have combining powers of 2.
   (b) HNO_3. (1) Hydrogen and the nitrate radical both have combining powers of 1.
   (c) $Fe(OH)_2$. (1) Iron has a combining power of 2; hydroxide has a combining power of 1. A bracket must be used for the OH.
   (d) $Al_2(CO_3)_3$. (1) Aluminium has a combining power of 3; carbonate has a combining power of 2. A bracket has to be used around the carbonate radical.

5 (a) $2Ca + O_2 \rightarrow 2CaO$. (1) Remember there must be the same number of atoms of calcium and oxygen on both sides of the equation. Balancing numbers can only be put in front of the symbols.
   (b) $N_2 + 3H_2 \rightarrow 2NH_3$. (1) There are 2 N atoms and 6 H atoms on each side of the equation.

6 (a) $Mg + 2HCl \rightarrow MgCl_2 + H_2$. (1)
   (b) $CH_4 + 2O_2 \rightarrow CO_2 + 2H_2O$. (1) Remember that if the chemical formulae are incorrect the equation will be as well.

7 $Pb^{2+}(aq) + 2OH^-(aq) \rightarrow Pb(OH)_2(s)$ Correct ion charges (1); correct balancing (1).

8 (a) $Al^{3+}(l) + 3e^- \rightarrow Al(s)$. (2) Positive ions gain electrons.
   (b) $2Cl^-(aq) \rightarrow Cl_2(g) + 2e^-$. (2) Negative ions lose electrons.

TOTAL

Solids, Liquids and Gases

● Solid, liquid and gas are the three states of matter. Changes of state are often brought about by changes in temperature:

$$\text{SOLID} \xrightleftharpoons[\text{freezing}]{\text{melting}} \text{LIQUID} \xrightleftharpoons[\text{condensing}]{\text{boiling}} \text{GAS}$$

● The three states of matter can be described using a simple particle model.

	Solid	Liquid	Gas
Arrangement	Regular	Random	Random
Distance between particles	Particles close together	Particles quite close together	Particles far apart
Forces between particles	Strong	Quite strong	Very weak
Movement of particles	Particles vibrate about a fixed point	Particles move randomly and quite slowly	Particles move randomly and quickly

● When a solid is heated the particles start to vibrate more vigorously. At the melting point the vibration is sufficient to break the forces holding the particles together and the particles start to move more freely. As the liquid reaches its boiling point the particles gain enough energy to break away from the surface.

● As a gas cools the particles move more slowly. Eventually the forces of attraction between the particles draw them closer together and a liquid forms. If the liquid is cooled further, the particles move even more slowly and eventually the forces of attraction bring the particles into a pattern and a solid forms.

Atomic Structure

- All atoms (except the simplest hydrogen atom) are made up of **protons**, **neutrons** and **electrons**. The protons and neutrons are found in the **nucleus** of the atom and the electrons are arranged in 'shells' around the nucleus.

- The atomic structure of an atom can be described using two numbers: the **atomic number**, the number of protons (or the number of electrons); and the **mass number**, the number of protons and neutrons.

MASS NUMBER
(the number of
protons + neutrons) — $_A^M$X — symbol for the element

ATOMIC NUMBER
(the number of protons which equals
the number of electrons)

- Atoms of the same element with different numbers of neutrons are called **isotopes**. For example, there are three isotopes of hydrogen.

Isotope	Symbol	Number of neutrons
Hydrogen	$_1^1H$	0
Deuterium	$_1^2H$	1
Tritium	$_1^3H$	2

- The electrons are arranged in shells around the nucleus. The shell can only accommodate a limited number of electrons, as shown in the table. Aluminium has an atomic number of 13 and so has 13 electrons. These are

Shell	Maximum no. of electrons
1	2
2	8
3	8

arranged with 2 in the first shell, 8 in the second shell and 3 in the third shell. This arrangement is written as 2, 8, 3.

- The structures of atoms can be shown very simply in an atom diagram.

$_6^{12}$C
Carbon

$_{16}^{32}$S
Sulphur

Chemical Bonding

● In chemical reactions atoms bond (join) together as a result of an interaction between the electrons of the atoms. There are two kinds of chemical bonding, **ionic** and **covalent**.

Ionic bonding	Covalent bonding
Involves electron transfer	Involves electron sharing
Occurs between a metal atom and a non-metal atom	Occurs between a non-metal atom and a non-metal atom
Metals lose electrons and form positive ions	Results in the formation of a molecule
Non-metals gain electrons and form negative ions	
Both atoms achieve complete outer electron shells	Both atoms achieve complete outer electron shells

● 'Dot and cross' diagrams can be used to show how the different types of bonds are formed.

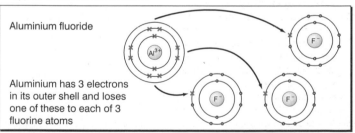

Aluminium fluoride

Aluminium has 3 electrons in its outer shell and loses one of these to each of 3 fluorine atoms

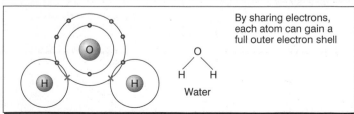

By sharing electrons, each atom can gain a full outer electron shell

O
H   H
Water

Structures and Properties

- Ionic compounds such as sodium chloride exist as **giant ionic lattices**. The ions are held firmly in place by strong electrostatic forces. These strong forces are the reason for the high melting and boiling points of ionic structures.

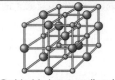

● chloride ion ○ sodium ion
The sodium chloride lattice

- Covalent compounds also contain very strong forces due to the covalent bonds. The forces within the molecules (**intramolecular** forces) are strong. The forces between the molecules (**intermolecular** forces) are often less strong. In graphite, for example, the forces between carbon atoms in the layers are strong, whereas the forces between the layers are much weaker. This explains why graphite is flaky and yet has a very high melting point.

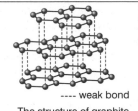

---- weak bond
The structure of graphite

Diamond is a giant atomic structure. Each atom is linked strongly to four other atoms

—— strong bond

The covalent bonds within the water molecule are strong but the bonds or forces between the molecules are relatively weak. This explains why water has a low boiling point.

Structure and Bonding (1–4)

1 What is the name of the process that describes the change from a solid to a liquid? (1)

2 In which state of matter are the particles only able to vibrate? (1)

3 What is the name of the number that indicates how many protons there are in an atom? (1)

4 Why are atoms neutral? (1)

5 What name is given to a charged atom or group of atoms? (1)

6 How many protons, neutrons and electrons are there in the phosphorus atom shown. (1)

> $^{31}_{15}$ P

7 What is the arrangement of the electrons in the phosphorus atom? (1)

8 Draw an atom diagram for this phosphorus atom. (1)

9 What name is given to the forces that exist between molecules? (1)

10 What type of chemical bonding occurs when calcium reacts with oxygen? Explain your answer. (1)

11 Draw a 'dot and cross' diagram to show the bonding when sodium combines with oxygen. (3) (atomic numbers: O = 8; Na = 11)

12 Draw a 'dot and cross' diagram to show the bonding in hydrogen chloride. (3) (atomic numbers: H = 1; Cl = 17)

13 Explain why sodium chloride has a high melting point. (2)

14 Look at the structures drawn for two different types of plastic, A and B. Which plastic would you expect to have the higher melting point? Explain your answer. (2)

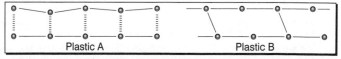

Plastic A   Plastic B

1 Melting. (1)
This change occurs at the melting point.

2 Solid. (1) In both liquids and gases the particles are able to move and do not vibrate about a fixed position.

3 Atomic number. (1)
This also gives the number of electrons in the atom.

4 Atoms contain equal numbers of protons and electrons. (1)
A proton has a single positive charge; an electron has a single negative charge.

5 An ion. (1) It is charged because it has different numbers of protons and electrons.

6 15 protons; 15 electrons; 16 neutrons. (1)
The lower number is the atomic number (p = e); the higher number is the mass number (p + n).

7 2, 8, 5. (1)

8 (1)

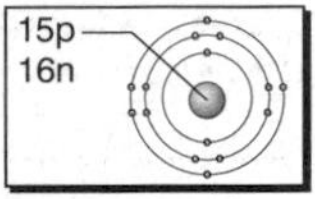

15p
16n

9 Intermolecular. (1)
Those that exist within a molecule are intramolecular.

10 Ionic. Calcium is a metal and oxygen a non-metal. (1)

11 (3)

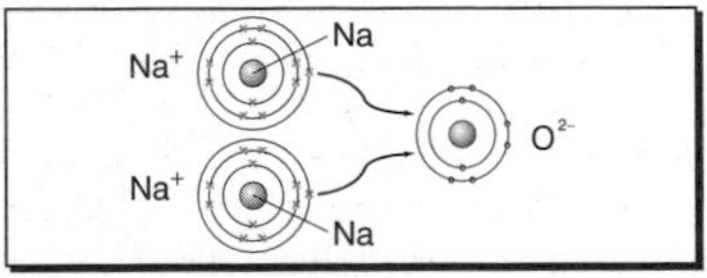

Na^+ Na
Na^+ O^{2-}
         Na

12 (3)

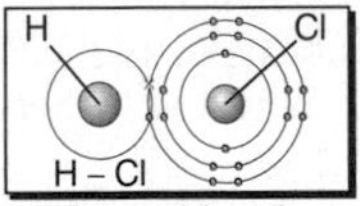

H   Cl
H – Cl

13 Sodium chloride contains ions (1) that are held together by strong electrostatic forces (1).

14 Plastic B will have the higher melting point. (1) There are strong intermolecular forces between the polymer chains. (1)

TOTAL

FUELS AND ENERGY (1)

Chemicals from Oil

- Crude oil was formed when the remains of dead sea animals were compressed by layers of rock over a period of millions of years.
- Crude oil is a **non-renewable** fuel because its supplies are limited and it will take millions of years to replace what we are currently using.
- Useful products such as petrol, diesel and paraffin can be obtained from crude oil by **fractional distillation**. The crude oil is heated in a furnace and passed into a large fractionating tower.

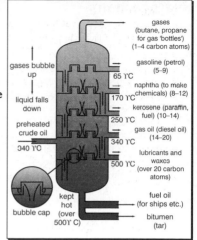

The fractions with the lower boiling points (smaller molecules) condense and are collected higher up the tower than those with higher boiling points (larger molecules).

- The quantities of the fractions obtained from crude oil do not match the demand. The lower boiling point fractions (e.g. petrol) are in greater demand than the higher boiling point fractions (e.g. naphtha). In the **cracking** process, which involves heating at high temperature in the presence of a catalyst, the larger molecules are broken down into more useful smaller molecules.

$$C_{10}H_{22} \text{ (l)} \rightarrow C_4H_{10} \text{ (g)} + 2C_3H_6 \text{ (g)}$$
decane → butane + propene

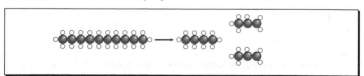

49

Hydrocarbons

- Crude oil is a mixture of **hydrocarbons**. Hydrocarbons are molecules that contain carbon and hydrogen atoms *only*.
- When a hydrocarbon burns in a plentiful supply of air or oxygen, carbon dioxide and water are formed. When the oxygen supply is limited, carbon (soot) or carbon monoxide are formed.

  methane + oxygen → carbon dioxide + water

  $CH_4(g) + 2O_2(g) \rightarrow CO_2(g) + 2H_2O(l)$
- Sulphur is an impurity in many hydrocarbon fuels. When these fuels are burnt sulphur dioxide is produced, which is one of the causes of **acid rain** (it reacts with rain water to form sulphuric acid).
- The carbon dioxide produced when hydrocarbon fuels are burnt is thought to be one of the causes of the **greenhouse effect**. This is when the energy from the Sun is trapped by the Earth and not reflected back into space.
- There are two important groups or families of hydrocarbons, called the **alkanes** and the **alkenes**. The alkanes are **saturated** hydrocarbons (they contain only single C–C bonds); the alkenes are **unsaturated** hydrocarbons (they contain at least one C=C bond).

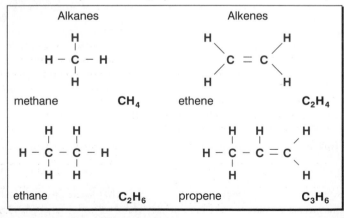

- Alkenes can be distinguished from alkanes by the fact that they decolorise bromine water:

$C_2H_4 + Br_2 = C_2H_4Br_2$
brown   colourless

- Alkenes can be used to make **addition polymers**, e.g. ethene (a **monomer**) can be converted into poly(ethene) (a **polymer**):

ethene   poly (ethene)

- Other polymers can be made in the same way.

vinyl chloride   poly (vinyl chloride)
(chloroethene)   [poly (chloroethene)]

Energy Transfers

- Energy transfers in chemical reactions can be categorised as **exothermic** or **endothermic**.

Type of reaction	Energy change	Temperature change of surroundings
Exothermic	Energy transferred to the surroundings	Temperature increases
Endothermic	Energy absorbed from the surroundings	Temperature decreases

FUELS AND ENERGY (4)

- Energy transfers are usually measured using the equipment shown below.

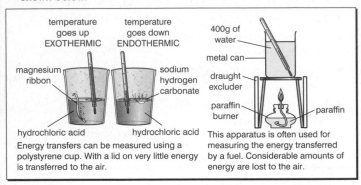

temperature goes up
EXOTHERMIC

temperature goes down
ENDOTHERMIC

magnesium ribbon

sodium hydrogen carbonate

hydrochloric acid

hydrochloric acid

Energy transfers can be measured using a polystyrene cup. With a lid on very little energy is transferred to the air.

400g of water

metal can

draught excluder

paraffin burner

paraffin

This apparatus is often used for measuring the energy transferred by a fuel. Considerable amounts of energy are lost to the air.

- The amount of energy transferred in a reaction can be calculated from the following formula:

energy = mass of liquid × specific heat capacity × temperature rise
(J)   (g)   (4.2 J/g/°C)   (°C)

- When chemical bonds are broken energy is absorbed (an endothermic process) and when chemical bonds are formed energy is released (an exothermic process). If a chemical reaction is exothermic overall, then more energy must be released on forming new bonds than is used up in breaking the old bonds. This balance can be seen in the following **energy level diagram**.

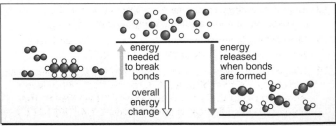

energy needed to break bonds

energy released when bonds are formed

overall energy change

Fuels and Energy (1–4)

1 What process is used to separate the components of crude oil? (1)

2 What two conditions are needed in the cracking process? (1)

3 What is a non-renewable fuel? (1)

4 What is a hydrocarbon? (1)

5 Are alkanes a family of saturated or unsaturated hydrocarbons? (1)

6 What chemical test can be used to distinguish an alkene from an alkane? (1)

7 What is the chemical formula of this hydrocarbon? (1)

$$H-\overset{\displaystyle \overset{H}{|}}{\underset{\displaystyle \underset{H}{|}}{C}}-\overset{\displaystyle \overset{H}{|}}{\underset{\displaystyle \underset{H}{|}}{C}}-\overset{\displaystyle \overset{H}{|}}{\underset{\displaystyle \underset{H}{|}}{C}}-H$$

8 Is the hydrocarbon in question 7 an alkane or an alkene? (1)

9 What is the name of the type of process used to make polyethene from ethene? (1)

10 What polymer is made using a monomer called vinyl chloride? (1)

11 How is crude oil made? (3)

12 Explain how the percentage of carbon dioxide in the atmosphere remains approximately constant. (3)

13 Explain how increasing amounts of carbon dioxide in the atmosphere could lead to global warming. (2)

14 1.5 g of octane was burnt in a spirit burner under a metal can containing 250 cm^3 of water. The temperature of the water increased by 30°C. Calculate the energy transferred to the water. (The specific heat capacity of water is 4.2 J/g/°C.) (2)

1 Fractional distillation. (1)
   In simple distillation the liquids would not be separated.
2 High temperature, catalyst. (1)
   The process is often called catalytic cracking.
3 One that takes a very long time to form (millions of years). (1)
   It is NOT one 'which cannot be used again'. No fuel can be
   used again once it has been burnt!
4 A molecule containing carbon and hydrogen atoms only. (1)
   You must include the word 'only'.
5 Saturated. (1)
   In a saturated molecule there are only C–C single bonds.
6 Add bromine water, if it decolorises it is an alkene. (1)
   Only unsaturated hydrocarbons will decolorise bromine.
7 C_3H_8 (1)
   This is propane.
8 Alkane. (1)
   Alkenes contain at least one C=C bond.
9 (Addition) polymerisation. (1)
   The key word is polymerisation.
10 Polyvinyl chloride (PVC). (1) The name of the polymer starts
   with 'poly' followed by the name of the monomer.
11 Dead sea creatures (1); compressed (1); over millions of
   years (1). Look for three separate points. Avoid saying
   'for a long time', fossil fuels form over millions of years.
12 Carbon dioxide is used up in photosynthesis and dissolves in
   the oceans (1); it is produced in respiration and combustion
   (1); the amount produced equals the amount used up (1).
   It is important to include the correct scientific terms (see
   Ecology and the Environment).
13 Carbon dioxide enables energy from the Sun to be trapped
   as heat instead of being reflected back into space (1) in a
   process known as the greenhouse effect (1). Remember that
   there is still disagreement as to whether global warming is
   caused by a greenhouse effect or natural climatic variation.
14 31 500 (1) J (1). Energy transferred = mass × specific heat
   capacity × temp rise; $E = 250 \times 4.2 \times 30 = 31\,500$ J.
   Don't forget the units.

TOTAL

Geological Change

- The Earth is made up of layers: a thin rocky **crust**; the **mantle** and the **core**. Radioactive decay in the centre of the Earth releases a lot of energy. This energy is sufficient to keep the outer core and the mantle in the liquid state. The outer crust (which is only between 5 and 70 km thick) is really floating on the liquid mantle. The liquid in the mantle is called **magma**.

- The crust is made up of large plates known as **tectonic plates**. Convection currents in the mantle cause these plates to move very slowly. If plates move together and collide, the more dense plate will be pushed down into the mantle (a process known as **subduction**) and an ocean trench will be formed. If two plates rub together an earthquake occurs. Where plates move apart a volcano or an ocean ridge will be formed.

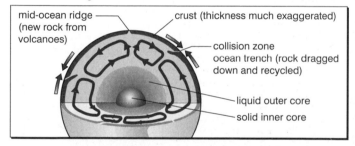

mid-ocean ridge
(new rock from
volcanoes)

crust (thickness much exaggerated)

collision zone
ocean trench (rock dragged
down and recycled)

liquid outer core

solid inner core

- The Earth's crust is made up of three types of rock.

Type	Method of formation	Appearance	Examples
Igneous	Cooling of hot magma	Hard, made up of crystals	Granite, basalt
Sedimentary	Layers of rock sediments, shells and bones are compressed	Layers, often containing fossils	Limestone sandstone
Metamorphic	Formed from igneous or sedimentary rocks by high temperature and pressure	Grains and crystals often distorted, fossils rare	Marble, slate

ROCKS AND METALS (2)

- Rocks are constantly being broken down and new rocks formed in a process called the **rock cycle**:
  1. igneous rocks are converted into sedimentary rocks by **weathering** and **erosion**
  2. sedimentary rocks are converted into metamorphic rocks by heat and pressure
  3. metamorphic rocks are converted into igneous rocks by being melted in the mantle and then reformed when they cool.

Reactivity Series of Metals

- Metals have very different reactivities. A reactivity series summarises the common reactions of common metals.
- The more reactive metals react with oxygen to form **oxides**.

  magnesium + oxygen → magnesium oxide

  $$2Mg(s) + O_2(g) \rightarrow 2MgO(s)$$

Metal	Reactivity	Reaction with air	Reaction with water	Reaction with hydrochloric acid
Potassium	Most		React with cold water to form a metal hydroxide and hydrogen	React violently to form a metal chloride and hydrogen
Sodium				
Calcium				
Magnesium		Burn in air or oxygen to form a metal oxide		
Aluminium			React with steam to form a metal oxide and hydrogen	React to form a metal chloride and hydrogen
Zinc				
Iron				
Tin		React with air and oxygen but do not burn		
Lead				
Copper			No reaction with water or steam	No reaction with dilute acid
Silver		Do not react with air or oxygen		
Gold	Least			

ROCKS AND METALS (3)

- The more reactive metals react with water or steam to form metal **hydroxides** or **oxides**.

  sodium + water → sodium hydroxide + hydrogen

  $2Na(s) + 2H_2O(l) \rightarrow 2NaOH(aq) + H_2(g)$

  magnesium + steam → magnesium oxide + hydrogen

  $Mg(s) + H_2O(g) \rightarrow MgO(s) + H_2(g)$

- The more reactive metals react with acids to form **salts** and hydrogen.

  zinc + hydrochloric acid → zinc chloride + hydrogen

  $Zn(s) + 2HCl(aq) \rightarrow ZnCl_2(aq) + H_2(g)$

  (note: hydrochloric acid always forms salts called *chlorides*)

  magnesium + sulphuric acid → magnesium sulphate + hydrogen

  $Mg(s) + H_2SO_4(aq) \rightarrow MgSO_4(aq) + H_2(g)$

  (note: sulphuric acid always forms salts called *sulphates*)

- A more reactive metal will **displace** a less reactive metal from a solution of one of its salts.

  zinc + copper(II) chloride → zinc chloride + copper

  $Zn(s) + CuCl_2(aq) \rightarrow ZnCl_2(aq) + Cu(s)$

- A simple electrical cell can be made by putting two different metals into a conducting solution or **electrolyte**. The voltage of the cell depends on the difference in reactivity of the two metals. The greater the difference in reactivity the greater the voltage.

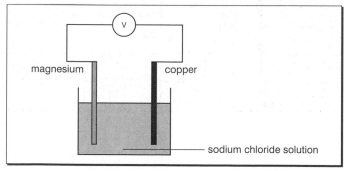

magnesium   copper

sodium chloride solution

Extraction of Metals

● Some unreactive metals, e.g. silver and gold, are found in the Earth's crust in their **native** state, i.e. uncombined with other elements. However, most metals are found in minerals chemically combined with other elements, i.e. as **compounds**. Chemical methods have to be used to separate the metal from the other elements. The method chosen depends on the reactivity of the metal:

Metal	Extraction method
Potassium	The most reactive metals are obtained using eletrolysis
Sodium	
Calcium	
Magnesium	
Aluminium	
(Carbon)	
Zinc	These metals are below carbon in the reactivity series and so can be obtained by heating their oxides with carbon
Iron	
Tin	
Lead	
Copper	
Silver	The least reactive metals are found as pure elements
Gold	

  1 the most reactive metals are obtained by **electrolysis**
  2 other metals are obtained by heating the mineral with **carbon**.
● In electrolysis the mineral is used as an **electrolyte**, usually in the molten state. A high current is passed through the electrolyte and the metal is formed at the negative **electrode** or **cathode**, e.g. aluminium is obtained from the mineral bauxite which is aluminium oxide. Aluminium ions move to the cathode and are deposited.
$$Al^{3+}(l) + 3e^- \rightarrow Al(s)$$
Oxygen gas is formed at the other electrode, the **anode**.
$$2O^{2-}(l) \rightarrow O_2(g) + 4e^-$$
● Carbon can only be used to extract metals that are below carbon in the reactivity series. On heating the carbon **reduces** the metal oxide to the metal, e.g. iron is obtained from the mineral haematite which is iron oxide.

iron(III) oxide + carbon → iron + carbon monoxide
$$Fe_2O_3(s) + 3C(s) \rightarrow 2Fe(s) + 3CO(g)$$

Rocks and Metals (1–4)

1 Look at the diagram showing the structure of the Earth. Label the parts named A, B and C. (3)

2 When two tectonic plates collide one of the plates is pushed into the mantle. What name is given to this process? (1) What determines which of the two plates is pushed into the mantle? (1)

3 Igneous and sedimentary are two types of rock. What is the name given to the other major type of rock? (1)

4 Aluminium is obtained from its mineral bauxite (aluminium oxide) using a process involving electricity.
   (a) What is the name of this type of process? (1)
   (b) Which electrode does the aluminium form at? (1)
   (c) What is the other product produced in the process? (1)
   (d) Why does the bauxite have to be in the molten (liquid) state? (1)
   (e) Write down an ionic equation to show the formation of aluminium in this process. (1)

5 The rock cycle can be used to show how the three main types of rock are linked together. Explain how new types of rock are formed as part of the rock cycle. (3)

6 This question is about the reactivity series of metals shown.
   (a) Name a metal that is found in its native state. (1)
   (b) (i) What process could be used to obtain lead from a mineral containing lead(II) oxide? (1)
       (ii) Write a symbol equation for this reaction. (1)
   (c) Potassium reacts with cold water. Write an equation for this reaction.(1)
   (d) (i) Write a word equation for the reaction of magnesium and hydrochloric acid. (1)
       (ii) Write the symbol equation for this reaction. (1)

| Potassium |
| Sodium |
| Calcium |
| Magnesium |
| Aluminium |
| Zinc |
| Iron |
| Lead |
| Copper |
| Silver |
| Gold |

1 A, Crust (1); B, Mantle (1); C, Core (1).
   Sometimes the core is shown divided into inner and outer.
2 Subduction. (1) The denser plate. (1)
   This occurs when an oceanic plate (more dense) collides
   with a continental plate (less dense).
3 Metamorphic. (1)
   Metamorphic rocks are formed when the other two types
   of rock are changed by heat and pressure.
4 (a) Electrolysis. (1) Heating with carbon will not work.
   (b) Cathode. (1) The aluminium ions have a positive charge.
   (c) Oxygen. (1) The oxide ions are attracted to the anode.
   (d) The ions must be free to move. (1) Electrolytes only
        conduct electricity when molten or when dissolved
        in water.
   (e) $Al^{3+}(l) + 3e^- \rightarrow Al$. (1) The ion gains three electrons to
        form the neutral atom.
5 Igneous rocks are broken into small pieces by weathering and
   erosion. These pieces are transported by water into streams
   and rivers and compressed to form sedimentary rocks. (1)
   These sedimentary rocks, when subjected to intense heat and
   pressure, are slowly changed into metamorphic rocks. (1)
   Subduction causes rocks to melt and form magma which
   forms igneous rock when it is expelled from volcanoes. (1)
   This is a very common question. Check you know the
   three stages.
6 (a) Silver or gold. (1) These are very unreactive metals.
   (b) (i) Heating with carbon. (1) Cheaper than electrolysis!
        (ii) $PbO(s) + C(s) \rightarrow Pb(s) + CO(g)$. (1)
             Or $2PbO(s) + C(s) \rightarrow 2Pb(s) + CO_2(g)$
   (c) $2K(s) + 2H_2O(l) \rightarrow 2KOH(aq) + H_2(g)$. (1) The most
        reactive metals form hydroxides (alkalis).
   (d) (i) magnesium + hydrochloric acid → magnesium
             chloride + hydrogen. (1) Hydrochloric acid forms
             salts called chlorides.
        (ii) $Mg(s) + 2HCl(aq) \rightarrow MgCl_2(aq) + H_2(g)$. (1) If you got
             the formula of magnesium chloride wrong check
             Chemical Formulae and Equations again.

TOTAL

CHEMICAL REACTIONS (1)

Collision Theory

- For a chemical reaction to occur the reacting particles (atoms, molecules or ions) must first collide. There must also be enough energy available in the collision so that chemical bonds can be broken. Not all collisions lead to a reaction. A collision that does have enough energy and leads to a reaction is called an **effective collision** or **successful collision**.

- Reactions occur at very different speeds. Explosions are very rapid reactions; rusting is a much slower reaction. All reactions have an energy barrier or **activation energy**. When particles collide there must be sufficient energy to exceed the activation energy if a reaction is to occur. Slow reactions generally have high activation energies.

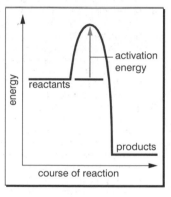

- A quick reaction takes place in a short time and so has a high rate. Care is needed not to confuse **time** and **rate**.

- The rate of many common reactions can be studied by measuring the loss in mass of the reactants or the volume of gas produced against time. A typical graph of the results of such a reaction is shown below.

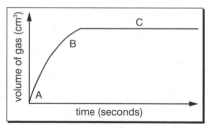

Where the rate of the reaction is at its greatest, the curve is at its steepest (A). The reaction becomes slower as it proceeds (B). When the reaction is complete the curve is horizontal (C).

Controlling the Rate of a Reaction

Factor	Effect on rate	Explanation
Concentration	Increasing the concentration of a reactant increases the rate.	Increasing the concentration results in the particles being more crowded. There will therefore be more frequent collisions and so more effective collisions.
Temperature	Increasing the temperature of a reaction increases the rate.	Increasing the temperature increases the kinetic energy of the particles. They will therefore move more quickly, collide more often and more energy will be available to overcome the activation energy.
Surface area	Increasing the surface area of a solid (making the particles smaller) increases the rate.	Increasing the surface area exposes more solid particles. There will therefore be more frequent collisions and so more effective collisions in a certain time.
Catalyst	Adding a catalyst changes the rate of a reaction – it usually increases it.	Catalysts change the activation energy of the reaction. Most reduce the activation energy which means that more collisions are going to have enough energy to exceed the activation energy.
Pressure	Increasing the pressure of reactions involving gases increases the rate.	The particles are pushed into a smaller volume and so collide more frequently. There will therefore be more effective collisions in a fixed time.
Light	Increasing the light intensity increases the rate of light-sensitive reactions.	Some particles are able to absorb energy from the light. As they then have more energy more collisions are likely to be effective.

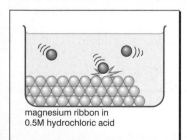

magnesium ribbon in
0.5M hydrochloric acid

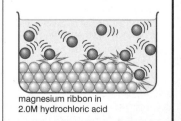

magnesium ribbon in
2.0M hydrochloric acid

CHEMICAL REACTIONS (3)

● Changing the rate of a reaction will not change the *amount* of product that is formed. However, it will change *how quickly* a certain amount of product is formed. On the graph, carbon dioxide is produced more quickly at 30°C than at 20°C but the total amount produced in the reaction (0.4 g) is the same. This is because there are the same amounts of reactants at both temperatures.

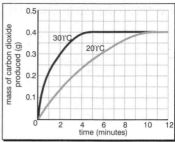

● Many of the chemical reactions in livings things are affected by **enzymes**. Enzymes are biological catalysts and are protein molecules. Enzymes often only catalyse one reaction: they are very specific. This is because they have an **active site** (where the reaction occurs) that is only the right shape for one type of molecule, i.e. only one type of molecule can fit in.

● Increasing the temperature usually increases the rate of a reaction, but not in reactions where an enzyme is involved. The protein structure of an enzyme is affected by temperature. Above a certain temperature the protein becomes **denatured** and ceases to function as a catalyst. Enzymes have an optimum temperature at which their reactions have the maximum rate. They are also affected by pH so have an optimum pH.

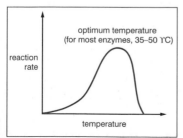

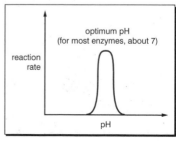

CHEMICAL REACTIONS (4)

Reversible Reactions

● In some chemical reactions, as well as being able to convert the reactants into the products, the products can be converted back into the reactants. Reactions of this type are **reversible.** For example, when copper(II) sulphate crystals are heated they form anhydrous copper(II) sulphate (a grey powder) and steam. If water is added to the grey powder the blue copper(II) sulphate is reformed.

$$CuSO_4.5H_2O(s) \rightleftharpoons CuSO_4(s) + 5H_2O(l)$$
blue crystals   grey powder

● Some important industrial processes are reversible. For example:

Production of ethanol

ethene + steam $\rightleftharpoons$ ethanol
$$C_2H_4(g) + H_2O(g) \rightleftharpoons C_2H_5OH(g)$$

Production of ammonia (the Haber Process)

nitrogen + hydrogen $\rightleftharpoons$ ammonia
$$N_2(g) \quad + \quad 3H_2(g) \quad \rightleftharpoons \quad 2NH_3(g)$$

● In a reversible reaction a point is reached when the rate of the reaction of reactants to products equals the rate of the reaction from products to reactants. In this situation the reaction is in **equilibrium.**
● The position of an equilibrium can often be changed by:
  ■ changing the concentrations of reactants and products
  ■ changing the pressure
  ■ changing the temperature.
● In industrial processes the conditions chosen often show a balance between getting as high a proportion as possible of products to reactants, while making sure the reaction proceeds at a fast enough rate. For example, in the Haber process the conversion of nitrogen and hydrogen to ammonia is favoured by low temperature. However, at low temperature the rate of reaction is slow. A compromise temperature of 450°C is used.

Chemical Reactions (1–4)

1 What name is given to the energy barrier that must be overcome before a reaction between two colliding particles can occur? (1)

2 Reaction A is completed in 2 minutes whereas reaction B takes 4 minutes. Which reaction has the greater rate of reaction? (1)

3 What is a catalyst? (1)

4 What is an enzyme? (1)

5 (a) Which has the greater surface area, 10 g of marble chips or 10 g of powdered marble? (1)
   (b) Which would react quickest with hydrochloric acid? (1)

6 What type of chemical reaction will have its rate affected by changing the pressure? (1)

7 What name is given to a reaction in which reactants form products and products change back to reactants? (1)

8 Look at the graph which shows the volume of hydrogen produced when magnesium ribbon is added to hydrochloric acid. Use the letters A, B and C to answer the following questions.

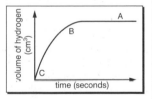

   (a) Where was the rate of reaction the greatest? (1)
   (b) Where was the rate of reaction zero? (1)

9 Explain why increasing temperature increases the rate of most reactions. (3)

10 Explain how a catalyst can increase the rate of a reaction. (2)

11 Why does an enzyme have an optimum temperature? (3)

12 In the Haber process for the manufacture of ammonia the conversion of nitrogen and hydrogen into ammonia is favoured by low temperature. Explain why a temperature as high as 450 °C is chosen. (2)

1 Activation energy. (1) The collision will only be successful if there is sufficient energy available to break chemical bonds.

2 A. (1) Try not to confuse rate and time. Reactions with a high rate take place in a short time.

3 A substance that changes the rate of a chemical reaction. (1) The catalyst is not chemically changed in the reaction.

4 An enzyme is a biological catalyst. (1)
Enzymes are involved in many of the reactions that take place in living cells.

5 (a) 10 g of powdered marble. (1)
(b) 10 g of powdered marble. (1)
The greater the surface area the greater the frequency of collisions.

6 A reaction involving gases. (1) Pressure will have little effect on reactions between solids and liquids.

7 A reversible reaction. (1)

8 (a) C. (1) The rate is greatest at the beginning of the reaction when there are more particles available to react.
(b) A. (1) No more gas is being produced, the reaction is complete.

9 The particles will be moving faster (1) and will therefore collide more frequently (1). The particles will also have more energy when they collide and so the energy of collision is more likely to exceed the activation energy. (1)
In a question worth 3 marks look for 3 key points.

10 A catalyst lowers the activation energy (1) so more collisions will have sufficient energy to lead to reactions (1).
A catalyst provides an alternative route or mechanism for the reaction.

11 On increasing temperature, initially the rate will increase (1) but eventually the enzyme will denature (1). The temperature at which the rate is at a maximum is the optimum temperature. (1)

12 The equilibrium is favoured by low temperature (1) but a higher temperature is needed to give a satisfactory rate of reaction (1). The conditions chosen are a balance between rate and equilibrium.

TOTAL

General Features

● Elements are arranged in a periodic table so that elements with similar properties and reactions are close together.

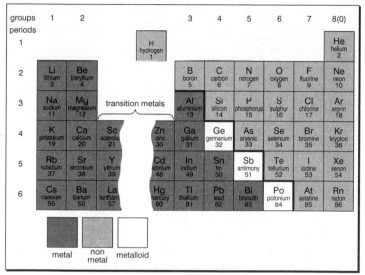

● The rows of elements are called **periods**.
● The columns of elements are called **groups**.
● Most of the elements are **metals**; the 'staircase' separates the metals from the **non-metals**.
● Some elements which have properties similar to both metals and non-metals, known as **metalloids**, are found close to the staircase, e.g. germanium, Ge.
● The middle block of elements are known as the **transition metals**.
● The numbers shown on the periodic table are the **atomic numbers**. This gives the number of protons (and electrons) in an atom of the element.

Metals and Non-Metals

● Metals and non-metals have quite different physical and chemical properties.

Typical physical properties of metals	Typical physical properties of non-metals
Good conductors of electricity	Poor conductors of electricity
Good conductors of heat	Poor conductors of heat
High melting points	Low melting points
Shiny	Dull
Hard Malleable – can be beaten into shape Ductile – can be drawn into wires Sonorous – rings when struck	Brittle

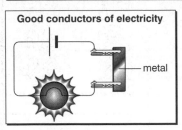

Good conductors of electricity

metal

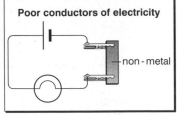

Poor conductors of electricity

non-metal

● Many metals and non-metals readily react with oxygen to form oxides. These oxides have very different chemical properties.

	Metal oxides	Non-metal oxides
Type of oxide	Basic	Acidic
pH of solution	More than 7 (alkaline)	Less than 7 (acidic)
Reactions	React with an acid to form a salt and water, e.g. $MgO(s) + 2HCl(aq) \rightarrow$ $MgCl_2(aq) + H_2O(l)$	React with an alkali to form a salt and water, e.g. $2NaOH(aq) + CO_2(g) \rightarrow$ $Na_2CO_3(aq) + H_2O(l)$

The Groups

● Elements in the same group have the same number of electrons in their outermost electron shells.

Group number	1	2	3	4	5	6	7	0
Electrons in outermost shell	1	2	3	4	5	6	7	2 or 8 (full)

● The most reactive elements are those which need to lose or gain one electron in order to obtain a full outer electron shell.

● Group 1 is known as the **alkali metals**. These are highly reactive metals. They are stored under oil to prevent reaction with air (oxygen) and water.

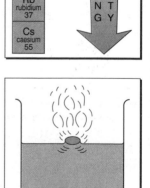

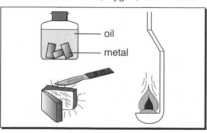

sodium + oxygen → sodium oxide
$4Na(s) + O_2(g) \rightarrow 2Na_2O(s)$

The metals react vigorously. They melt, float on the surface of the water and, in the case of the more reactive elements, the hydrogen burns. The hydroxide formed is an alkali.

sodium +   water   → sodium hydroxide + hydrogen
$2Na(s) + 2H_2O(l) \rightarrow 2NaOH(aq) + H_2(g)$

- Group 7 is known as the **halogens**. These are highly reactive non-metals and react with most metals. They exist as **diatomic** molecules, i.e. molecules containing two atoms, e.g. F_2, Cl_2.

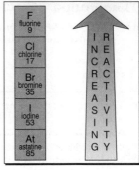

With metals, salts are formed.

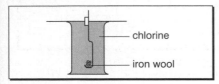

magnesium + chlorine → magnesium chloride

$$Mg(s) + Cl_2(g) \rightarrow MgCl_2(s)$$

The halogens undergo **displacement** reactions. A more reactive halogen will displace a less reactive halogen from a salt.

chlorine + potassium iodide → potassium chloride + iodine

$$Cl_2(aq) + 2KI(aq) \rightarrow 2KCl(aq) + I_2(aq)$$

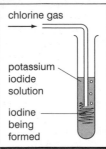

- Group 0 is known as the **noble gases**. As they have full outer electron shells they are very unreactive. Unlike the halogens they exist as single atoms, i.e. they are **monatomic**, e.g. He, Ne.

	Uses
He	Balloons
Ne	Red lights
Ar	Light bulbs

The Transition Metals

The transition metals are found in the centre of the periodic table. They are much less reactive than the alkali metals and so are more 'everyday' metals. They generally have high melting and boiling points and high densities. They are often used in industry as catalysts. Compounds of transition metals are usually coloured and so can easily be distinguished from those of other metals, which are usually white.

The Periodic Table (1–4)

1 Look at the diagram showing part of the periodic table.
The letters stand for elements.

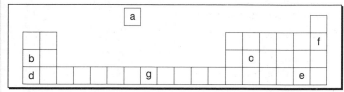

(a) Which element is in group 4? (1)
(b) Which element is in the second period? (1)
(c) Which elements are metals? (1)
(d) Which element is a metalloid? (1)
(e) Which element is a transition metal? (1)
(f) Which elements are gases? (1)
(g) Which element has an atomic number of 1? (1)
(h) Which element will be the more reactive, b or d?
Give a reason for your answer. (1)

2 Why do all the elements in group 1 have very similar chemical
properties? (1)

3 Phosphorus is an element in the third period in group 5.
Will phosphorus oxide be acidic or alkaline? Give a reason for
your answer. (1)

4 Potassium reacts with water. What would you SEE when
potassium is added to water? (2). Write a word equation for
the reaction. (2)

5 Write a word equation for the reaction between iron and chlorine
to form iron(III) chloride. (1). Write a symbol equation for this
reaction. (2)

6 The halogens undergo displacement reactions. What is a
displacement reaction? (1) Write a word equation for a reaction
in which chlorine displaces bromine. (2)

1 (a) c. (1) The groups are the columns.
 (b) f. (1) The periods are the rows.
 (c) b, d, g. (1) The metals are on the left side of the 'staircase'.
 (d) c. (1) This element is next to the 'staircase'.
 (e) g. (1) The transition elements are found in the middle of the periodic table.
 (f) a, f. (1) Gases must be non-metals. The non-metals with the lowest atomic numbers are most likely to be gases.
 (g) a. (1) This is hydrogen, the first element in the periodic table.
 (h) d. (1) Reactivity increases as you go down group 1. Remember that the reverse is true in group 7.

2 They all contain the same number of electrons in their outermost electron shell. (1) In group 1 this number is 1.

3 Acidic. Phosphorus is a non-metal; non-metal oxides are acidic. (1) Remember, it is metal oxides that are alkaline.

4 Any two from: The metal floats on the surface of the water (1); it melts and forms a ball (1); it effervesces (fizzes) (1); it quickly dissolves (1); a flame is seen (1). A common mistake is to write about the names of the products (hydrogen and potassium hydroxide) rather than what you would SEE or observe.
potassium + water → potassium hydroxide (1) + hydrogen (1)
Remember that the alkalis produced by the alkali metals are hydroxides.

5 iron + chlorine → iron(III) chloride (1)
$2Fe(s) + 3Cl_2(g) \rightarrow 2FeCl_3(s)$ (1) for correct formulae; (1) for correct balancing. All symbol equations should be balanced.

6 A displacement reaction is one in which a more reactive element displaces (takes the place of) a less reactive element from a salt. (1)
chlorine + sodium bromide → sodium chloride + bromine
(1) for correct reactants; (1) for correct products.
Other salts could be used, e.g. potassium bromide.

TOTAL

CHEMICAL CALCULATIONS (1)

Atomic Masses

- Atomic masses are given on a **relative atomic mass scale**. The masses of all atoms are compared with that of a carbon atom.
- The relative atomic mass (R.A.M.) of an atom is defined as: the mass of an atom on a scale where the mass of a carbon atom is 12 units.
- Using the relative atomic mass scale you can see that:

1 atom of carbon is 12× the mass of 1 atom of hydrogen

1 atom of sulphur is 2× the mass of 1 atom of oxygen

Relative atomic masses	
H = 1	S = 32
C = 12	Ca = 40
O = 16	Fe = 56
Mg = 24	Cu = 64

The Mole

- The mole is a number approximately equal to 6×10^{23}. It is important because:

6×10^{23} atoms of carbon have a mass of 12 g

6×10^{23} atoms of sulphur have a mass of 32 g

The relative atomic mass of any atom when expressed in grams contains 6×10^{23} atoms (1 mole of atoms).

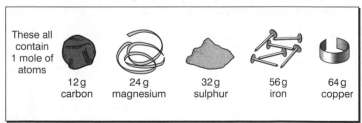

These all contain 1 mole of atoms

| 12 g carbon | 24 g magnesium | 32 g sulphur | 56 g iron | 64 g copper |

● Calculations can be carried out using the formula:

$$\text{moles of atoms} = \frac{\text{mass}}{\text{R.A.M.}}$$

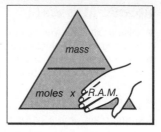

For example: How many moles of atoms are there in 80 g of calcium? (Ca = 40)

$$\text{moles} = \frac{80}{40} = 2$$

For example: What is the mass of 3 moles of carbon atoms? (C = 12)

mass = moles × R.A.M. = 3 × 12 = 36 g

Working Out Chemical Formulae

● This can be done by working out the number of moles of each atom present in a compound.

For example: 3 g of magnesium combines with 2 g of oxygen to make 5 g of magnesium oxide. What is the chemical formula of magnesium oxide? (O = 16; Mg = 24)

	Mg	**O**
Masses of elements	3	2
No. of moles	3/24 = 0.125	2/16 = 0.125
Simple ratio (divide by smallest)	0.125/0.125 = 1	0.125/0.125 = 1
Formula	MgO	

Moles of Molecules

● The mass of 1 mole (6×10^{23}) of molecules is the relative molecular mass (R.M.M.) expressed in grams.

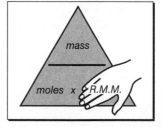

For example:
mass of 1 mole of carbon dioxide (CO_2) = 12 + 16 + 16 = 44 g

CHEMICAL CALCULATIONS (3)

Reacting Masses

● Chemical equations can be used to work out how the quantities of reactants and products are linked.
● In a balanced equation the numbers in front of each symbol or formula indicate the number of moles represented.

For example: Calcium burns in oxygen to form calcium oxide.
What mass of calcium oxide is made when 8 g of calcium is burnt?
($Ca = 40$; $O = 16$)

1 Write down the equation:	$2Ca(s)$	$+$	$O_2(g)$	$\rightarrow$ $2CaO(s)$
2 Write down the number of moles:	2 moles		1 mole	2 moles
3 Convert moles to masses:	80 g		32 g	112 g
4 Scale masses to those used:	8 g			11.2 g
	(scaling ÷ 10)			(scaling ÷ 10)

So, 8 g of calcium would produce 11.2 g of calcium oxide.

For example: Burning a sample of methane (CH_4) produced 88 g of carbon dioxide. What mass of methane was burnt?
($H = 1$; $C = 12$; $O = 16$)

1 Write down the equation:	$CH_4(g) +$	$2O_2(g) \rightarrow$	$CO_2(g) +$	$2H_2O(l)$
2 Write down the number of moles:	1 mole	2 moles	1 mole	2 moles
3 Convert moles to masses:	16 g	64 g	44 g	36 g
4 Scale masses to those used:	32 g		88 g	
	(scaling × 2)		(scaling × 2)	

So, 32 g of methane would be needed to produce 88 g of carbon dioxide.

Moles of Gases

● With a gas it is often more convenient to measure its volume rather than its mass.

1 mole of any gas occupies $22\,400\ cm^3$ at standard temperature and pressure (s.t.p. = 273 K and 101 325 N/m²)

1 mole of any gas occupies $24\,000\ cm^3$ (approx.) at room temperature and pressure (r.t.p.).

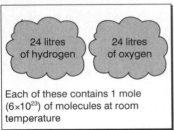

24 litres of hydrogen
24 litres of oxygen

Each of these contains 1 mole (6×10^{23}) of molecules at room temperature

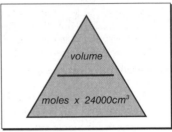

volume

moles x 24000cm³

For example: A container holds $12\,000\ cm^3$ of hydrogen at room temperature and pressure. How many moles of hydrogen does it contain?

$$moles = \frac{volume}{24\,000} = \frac{12\,000}{24\,000} = 0.5$$

Moles of Solutions

● A solution is made by dissolving a **solute** in a **solvent**.
● The concentration of a solution is usually expressed in terms of **molarity** or moles of solute per litre of solution.
● 1 mole of solute dissolved to make $1000\ cm^3$ of solution produces a 1 M (molar) solution.

For example: 0.5 mole of a solute is dissolved to make $250\ cm^3$ of solution. What is the solution's molarity?
0.5 mole in $250\ cm^3$ is the same concentration as 2 moles in $1000\ cm^3$. The concentration is 2M.

Chemical Calculations (1–4)

1 The relative atomic mass of oxygen is 16 and that of copper is 64. How many times heavier than an atom of oxygen is an atom of copper? (1)

2 Calculate the masses of the following:
   (a) 2 moles of carbon (C = 12). (1)
   (b) 10 moles of iron (Fe = 56). (1)
   (c) 0.25 moles of calcium (Ca = 40). (1)

3 How many moles of atoms are there in the following?
   (a) 23 g of sodium (Na = 23). (1)
   (b) 48 g of oxygen (O = 16). (1)
   (c) 1.4 g of nitrogen (N = 14). (1)

4 Calculate the relative molecular masses of the following:
   (a) H_2S (H = 1; S = 32). (1)
   (b) SO_2 (S = 32; O = 16). (1)
   (c) H_2SO_4 (H = 1; O = 16; S = 32). (1)

5 Calculate the masses of the following:
   (a) 10 moles of carbon dioxide, CO_2 (C = 12; O = 16). (1)
   (b) 0.5 moles of calcium carbonate, $CaCO_3$ (Ca = 40; C = 12; O = 16). (1)

6 Carbon burns in oxygen to form carbon dioxide. The equation for the reaction is: $C(s) + O_2(g) \rightarrow CO_2(g)$. Calculate the mass of carbon dioxide produced when 3 g of carbon is burnt. (C = 12; O = 16). (3)

7 Magnesium reacts with hydrochloric acid as shown in the equation: $Mg(s) + 2HCl(aq) \rightarrow MgCl_2(aq) + H_2(g)$
   (a) Calculate the mass of hydrogen produced from 2.4 g of magnesium (Mg = 24; H = 1). (3)
   (b) What volume would this mass of hydrogen occupy at room temperature and pressure? (molar volume = 24 000 cm^3 at r.t.p.) (2)

1 $4 \times$ heavier. (1) $64/16 = 4$.

2   **(a)** 24 g. (1) Mass = moles $\times$ R.A.M. = $2 \times 12 = 24$.
     (b) 560 g. (1) Mass = moles $\times$ R.A.M. = $10 \times 56 = 560$ g.
     (c) 10 g. (1) Mass = moles $\times$ R.A.M. = $0.25 \times 40 = 10$ g.

3   **(a)** 1 mole. (1) Moles = mass/ R.A.M. = $23/23 = 1$.
     (b) 3 moles. (1) Moles = mass/R.A.M. = $48/16 = 3$.
     (c) 0.1 moles. (1) Moles = mass/R.A.M. = $1.4/14 = 0.1$.

4   **(a)** 34. (1) R.M.M. = $1 + 1 + 32 = 34$. Remember that the
         R.M.M. has no units.
     (b) 64. (1) R.M.M. = $32 + 16 + 16 = 64$.
     (c) 98. (1) R.M.M. = $1 + 1 + 32 + 4(16) = 34 + 64 = 98$.

5   **(a)** 440 g. (1) R.M.M. = $12 + 2(16) = 44$.
        Mass = moles $\times$ R.M.M. = $10 \times 44 = 440$ g.
     (b) 50 g. (1) R.M.M. = $40 + 12 + 3(16) = 52 + 48 = 100$.
        Mass = moles $\times$ R.M.M. = $0.5 \times 100 = 50$ g.

6

$C(s)$ +	$O_2(g) \rightarrow$	$CO_2(g)$	
1 mole	1 mole	1 mole	(1) Write down the number of moles
12 g		44 g	(1) Convert moles to masses
3 g		11 g	(1) Scale masses
(scaling $\div 4$)		(scaling $\div 4$)	

7   **(a)**

$Mg(s)$ +	$2HCl(aq) \rightarrow$	$MgCl_2(aq)$ +	$H_2(g)$	
1 mole	2 moles	1 mole	1 mole	(1) moles
24 g			2 g	(1) masses
2.4 g			0.2 g	(1) scale
(scaling $\div 10$)			(scaling $\div 10$)	

     (b) 0.1 mole of hydrogen is produced. (1)
        Volume = $0.1 \times 24\,000 = 2400 \text{ cm}^3$ at r.t.p. (1)

Note: In questions 6 and 7 it is only necessary to convert moles into mass for those parts of the equation included specifically in the question. In question 7 there is no need to work out the mass of the hydrochloric acid.

                                                TOTAL

Electrical Circuits

- In the simple circuit shown, the battery 'pushes' electrical charge round the circuit to make a **current**. The battery also transfers energy to the electrical charge. The **voltage** is a measure of how much push the battery can provide and how much energy it can transfer to the charge.

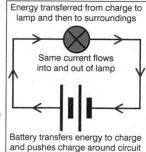

Energy transferred from charge to lamp and then to surroundings

Same current flows into and out of lamp

Battery transfers energy to charge and pushes charge around circuit

- An electric current is actually a flow of electrons. The electrons flow from negative to positive. However, circuit diagrams (like that shown) always show the current flowing from positive to negative. This is known as **conventional current**.

- There are two types of electrical circuit, **series** and **parallel**. In a series circuit there is only one route round the circuit, whereas in a parallel circuit there are at least two routes. In the parallel circuit the battery can push the charge along two alternative paths so more charge can flow round the circuit each second. The lamps in the parallel circuit shown will therefore be brighter than the lamps in the series circuit.

$$\text{current (I)} = \frac{\text{charge (Q)}}{\text{time (t)}} \qquad I = \frac{Q}{t}$$

Current is measured in amps (A), charge in coulombs (C) and time in seconds (s).

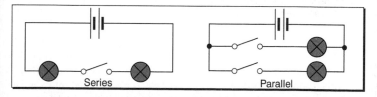

Series

Parallel

- As charge flows round a circuit it transfers energy to the components. The amount of energy that a charge transfers between two points is called the **potential difference** (**p.d.**). The p.d. is measured in volts.

- In a circuit the current is measured using an **ammeter** and the p.d. is measured using a **voltmeter**.

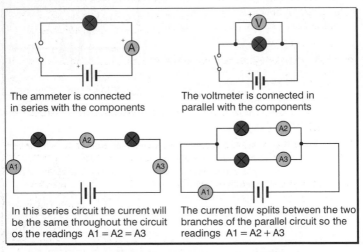

The ammeter is connected in series with the components

The voltmeter is connected in parallel with the components

In this series circuit the current will be the same throughout the circuit os the readings A1 = A2 = A3

The current flow splits between the two branches of the parallel circuit so the readings A1 = A2 + A3

- A battery produces a steady current known as a **direct current** (**d.c.**). Mains electricity is different, the direction of the current changes 50 times each second. This is known as **alternating current** (**a.c.**).

Resistance

- Substances that allow an electric current to flow through them are called **conductors**. Substances that do not are called **insulators**. In some conductors the electrons are able to flow more freely than in other conductors. These conductors have a lower **resistance**.

ELECTRICITY (3)

- Metals are good conductors because of their structure. Metal ions are surrounded by a 'sea' of electrons. When a p.d. is applied to the metal the electrons are able to

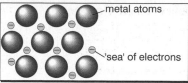

metal atoms

'sea' of electrons

move. They bump into the metal ions causing resistance.

- The greater the length of a conductor the greater its resistance. However, the greater the cross-sectional area of a conductor the lower its resistance.

- The relationship between voltage, current and resistance is given by **Ohm's law**:

voltage (V) = current (I) × resistance (R)   V = IR

Voltage is measured in volts (V), current in amps (A) and resistance in ohms (Ω).

For example: Calculate the resistance of a heater element if the current is 20 A when it is connected to a 230 V supply.

Write down the formula in terms of R: $R = \dfrac{V}{I}$

Substitute the values for V and I: $R = \dfrac{230}{20}$

Write down the answer and unit: R = 11.5 ohms

cover I to find $I = \dfrac{V}{R}$

- The resistance of most conductors increases with temperature: as the temperature increases the metal ions vibrate more and restrict the movement of the electrons.

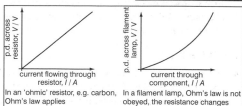

In an 'ohmic' resistor, e.g. carbon, Ohm's law applies

In a filament lamp, Ohm's law is not obeyed, the resistance changes with temperature

Electrical Power

● All electrical equipment has a **power** rating in **watts**. The power rating indicates how many joules of energy are supplied each second.

power (P) = voltage (V) × current (I)
$$P = V\,I$$

Power is measured in watts (W), voltage in volts (V) and current in amps (A).

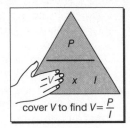

cover V to find $V = \dfrac{P}{I}$

For example: An electric kettle has a power rating of 2.3 kW (2300 W). What current will flow when the kettle is used on a 230 V supply?

Write down the formula in terms of I: $I = \dfrac{P}{V}$

Substitute the values: $I = \dfrac{2300}{230}$

Write down the answer and unit: $I = 10\,A$

● Electrical energy used at home is measured in kilowatt-hours (kWh). 1 kWh is the amount of energy transferred by a 1 kW device in 1 hour.

● Electrical appliances can be damaged if too high a current flows through them. **Fuses** are used to protect them. If the current gets too high the fuse wire melts and breaks the circuit.

● People using electrical devices are protected in two ways. Many devices are made of plastic: they are **double insulated**. If the live wire in the device becomes loose and touches the outer casing no current will flow because plastic is an insulator. When devices are made of metal an **earth wire** is used. If the live wire touches the casing the current will flow rapidly to earth. This flow of current would quickly melt the fuse. In this way the fuse and the earth wire work together to protect the user.

Electricity (1–4)

1 An electric current is a flow of what type of particles? (1)

2 Look at the electric circuit, which contains three identical lamps.

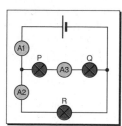

   (a) Give the letters of two lamps that are connected in series. (1)

   (b) Give the letters of two lamps that are connected in parallel. (1)

   (c) If lamp P was broken which, if any, of the remaining lamps would stay on? (1)

   (d) The reading on ammeter A1 is 0.6 A.

      (i) Which ammeter reading will be the higher, A2 or A3? Explain your answer. (2)

      (ii) What will be the currents shown on A2 and A3? (2)

3 What do the terms 'a.c.' and 'd.c.' mean? (2)

4 What term is used to describe a substance that does not conduct electricity? (1)

5 What will happen to the resistance of a piece of iron wire if:

   (a) its length is increased? (1)

   (b) its cross-sectional area is increased? (1)

6 A potential difference of 6 V is applied to a 60 | resistor. What current will flow through the resistor? (2)

7 What current flows through a kettle element with a power of 1150 W when it is connected to a 230 V supply? (2)

8 Explain how the fuse and the earth wire work together to make an electrical appliance such as a cooker safe to use. (3)

1 Electrons. (1)

2 **(a)** P and Q. (1)
These are on the same route round the circuit.
(b) P and R or Q and R. (1)
These are on different routes round the circuit.
(c) R. (1)
The current can still flow round this part of the circuit.
(d) **(i)** A2. (1) The resistance in this branch of the circuit will be lower as there is only one lamp. (1) Remember that the question states that the lamps are identical.
(ii) A2: 0.4 A. (1) A3: 0.2 A. (1) The resistance of P and Q will be double the resistance of R. Therefore the current through R will be double that through P and Q. The total current must be 0.6 A.

3 Alternating current (1) and direct current (1). Remember that mains electricity is a.c. whereas a battery provides d.c.

4 Insulator. (1)

5 **(a)** Resistance increases. (1) The longer the wire the harder it is for the electrons to 'fight' their way past the metal ions.
(b) Resistance decreases. (1) The greater the area the more routes there are for the electrons to travel along the wire. If in doubt look back at the diagram showing the structure of a metal.

6 $I = \dfrac{V}{R}$. (1) $I = \dfrac{6}{60} = 0.1$ A. (1)

This is Ohm's law. Don't forget to include the unit.

7 $I = \dfrac{P}{V}$. (1) $I = \dfrac{1150}{230} = 5$ A. (1)

You must be able to remember and rearrange this formula.

8 If the live wire became loose and touched the metal casing of the cooker the current would immediately flow to earth. (1) The route to the earth has very little resistance and so the current would increase. (1) This sudden increase in the current would melt or 'blow' the fuse. (1) Many electrical appliances have plastic outer cases. They are double insulated and do not need an earth wire.

TOTAL

Simple Magnetism

- The simple rule of magnetism is: opposite poles attract; like poles repel.

- A magnetic compass always points to the Earth's Magnetic North Pole. The end that does this is called the 'north-seeking pole' or **N-pole**. The other end is the 'south-seeking pole' or **S-pole**.

- Not all metals are attracted to a magnet. The only magnetic metals are iron, cobalt and nickel.

- The region around a magnet where its magnetic effect can be detected is called a **magnetic field**. The shape of a magnetic field can be worked out using iron filings or plotting compasses.

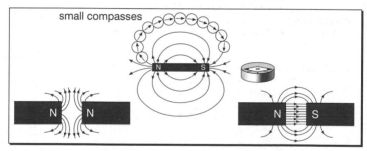

Electromagnetism

- When an electric current flows through a wire it creates a magnetic field. This is known as **electromagnetism**.

- The magnetic field is made stronger if the wire is made into a coil and is wrapped around a piece of magnetic material such as iron.

- The strength of an electromagnet can be increased by:
  - increasing the number of coils in the wire
  - increasing the current flowing through the wire.

ELECTROMAGNETISM (2)

A Relay

● A **relay** is like an electromagnetic switch. It uses a small current from a low-voltage circuit to switch on a higher current in a second higher-voltage circuit.

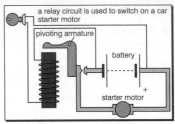

a relay circuit is used to switch on a car starter motor

pivoting armature

battery

starter motor

D.C. Motor

● The operation of an **electric motor** depends on the fact that when a current flows through the coil of wire it creates a magnetic field. This magnetic field then interacts with the magnetic field created by the permanent magnets and produces a force.

This rule shows the direction of the force

thuMb
Movement

First finger
Field

seCond finger
Current induced

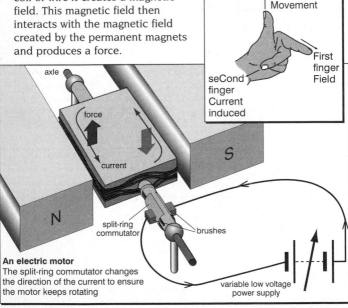

axle

force

current

N

S

split-ring commutator

brushes

variable low voltage power supply

An electric motor
The split-ring commutator changes the direction of the current to ensure the motor keeps rotating

Generators

- Electricity can be generated using a magnetic field involving a process known as **electromagnetic induction**. A current can be produced in a wire if:
  - the wire is moved through the magnetic field
  - the magnetic field is moved past the wire
  - the strength of the magnetic field around the wire is changed.

- A simple example of a current generator is a **dynamo**. A dynamo looks very like an electric motor. A split-ring commutator ensures that the current produced is a direct current. In contrast, power station **generators** produce an alternating current.

Generating Electricity

- The generation of electricity can be shown by the flow chart.

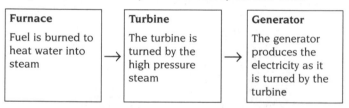

Furnace	**Turbine**	**Generator**
Fuel is burned to heat water into steam	The turbine is turned by the high pressure steam	The generator produces the electricity as it is turned by the turbine

- The electricity is transmitted round the country using the **National Grid**. To reduce the power loss in the grid the electricity is transmitted with a low current. (A high current would heat the transmission wire, causing a wasteful transfer of energy.)

- As the electricity is generated as a.c. it can be 'stepped up' and transmitted at high voltage (low current) using a transformer.

Transformers

- A transformer is a device for changing voltage. If it increases voltage it is known as a **step-up transformer**; if it decreases the voltage it is known as a **step-down transformer**.

- A transformer consists of two coils of insulated wire, wound on a piece of iron. With an alternating voltage across the primary coil, the alternating current creates a changing magnetic field in the core. This induces an alternating current in the second coil and hence an alternating voltage across this coil.

- If there are more turns on the secondary coil than on the primary coil the voltage will be stepped-up. The exact relationship is given by the equation:

$$\frac{\text{primary coil voltage (Vp)}}{\text{secondary coil voltage (Vs)}} = \frac{\text{number of primary turns (Np)}}{\text{number of secondary turns (Ns)}}$$

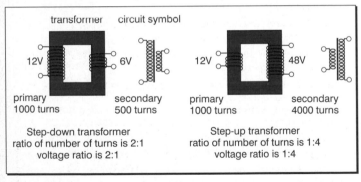

transformer   circuit symbol

12V — 6V

primary
1000 turns

secondary
500 turns

Step-down transformer
ratio of number of turns is 2:1
voltage ratio is 2:1

12V — 48V

primary
1000 turns

secondary
4000 turns

Step-up transformer
ratio of number of turns is 1:4
voltage ratio is 1:4

For example: Calculate the output voltage from a transformer when the input voltage is 230 V, the number of turns on the primary coil is 1000, and the number of turns on the secondary coil is 25.

$$\frac{Vp}{Vs} = \frac{Np}{Ns}; \frac{230}{Vs} = \frac{1000}{25}; Vs = 230 \times \frac{25}{1000} = 5.75\,V$$

Electromagnetism (1–4)

1 List two ways of increasing the strength of an electromagnet. (2)

2 Can an electromagnet be used to separate aluminium cans from other household waste? Explain your answer. (1)

3 What device uses a small current from a low-voltage circuit to switch on a higher current in a higher voltage circuit? (1)

4 What is the main difference between a d.c. motor and a dynamo? (1)

5 (a) What is the name given to the system of electrical cables that are used to transmit electricity round the country? (1)

   (b) Electricity is transmitted at very high voltages. Explain why high voltages are used. (2)

   (c) What device is used to change the voltage of the electricity produced in the power station to the value that it is transmitted at? (1)

6 (a) What are the main stages used in the production of electricity in a power station? (3)

   (b) Why is it important that the electricity is produced with an alternating current? (1)

7 A transformer with 10 turns in the primary coil and 30 turns in the secondary coil has an output voltage of 60 V. What will the input voltage be? (3)

8 Look at the diagram of an electric bell. Explain how the bell works when the switch is closed. (4)

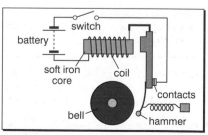

switch
battery
soft iron core
coil
contacts
bell
hammer

1 Increasing the number of coils (1); increasing the current (1). Increasing the voltage would also have the effect of increasing the current.

2 No. Aluminium is not magnetic. (1)
It is a common mistake to assume that all metals are magnetic.

3 A relay. (1)
The advantage in a car is that much thinner wires can be used for the relay circuit than are needed for the battery and starter motor circuit.

4 A motor uses electricity to produce motion (kinetic energy) whereas a dynamo uses motion to generate electricity. (1)
Motors and dynamos often look very similar.

5 (a) The National Grid. (1)
(b) There is less energy lost or wasted when transmitting at high voltage. (1) This is because the current is much smaller at higher voltages. (1)
(c) Transformer. (1)
It would actually be a step-up transformer. A step-down transformer is used near to homes to convert the voltage back to 230 V.

6 (a) Burning of fuel to change water into steam. (1) Use of high pressure steam to turn the turbines. (1) The turbines then turn the a.c. generators. (1)
In a nuclear power station the energy from the nuclear reaction is used to boil the water.
(b) Transformers only work with an a.c. (1)
A d.c. supply could not be stepped up or down.

7 $\dfrac{Vp}{Vs} = \dfrac{Np}{Ns}$. (1) $\dfrac{Vp}{60} = \dfrac{10}{30}$. (1) $Vp = 60 \times \dfrac{10}{30} = 20\,V$. (1)

It is always very important to show your working. Don't forget to give a unit in your answer.

8 When the switch is pressed the electromagnet attracts the hammer support and the hammer hits the bell. (1) The movement of the hammer support breaks the circuit (1) and so the electromagnet ceases to operate (1). The hammer support then returns to its original position forming the circuit again and the process is repeated. (1)

TOTAL

The Effect of Forces

● Forces can change the shape of an object, the speed of an object or its direction of movement. Forces are measured in **newtons**.

● Common forces include **friction**, which is the force that stops movement between touching surfaces, and **gravity**, which is the force caused by the attraction of objects to large masses (e.g. the Earth, Moon, etc.). **Weight** is another force and is different from **mass**.

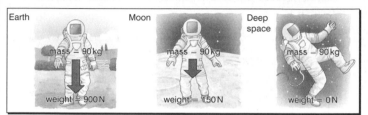

● **Hooke's law** shows the relationship between the force applied to an elastic material and the amount of stretching that is produced:

force (F) = constant (k) × extension (x) $F = k\,x$

● Usually there will be at least two forces acting on an object. If the forces are **balanced** the object will either be stationary or will be moving at constant speed. If the forces are **unbalanced** the object will be accelerating or decelerating.

● When the forces on a skydiver are balanced, as shown below, the **terminal speed** or **terminal velocity** will have been reached.

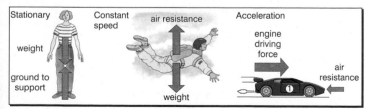

Speed, Velocity and Acceleration

● The **speed** of an object can be calculated as follows:

speed (v) = $\dfrac{\text{distance (s)}}{\text{time (t)}}$ $v = \dfrac{s}{t}$

Speed is measured in metres per second (m/s), distance in metres (m) and time in seconds (s).

● **Velocity** is almost the same as speed. It has a size (speed) and a direction.

● **Acceleration** is a measure of how much speed changes in a certain time:

acceleration (a) = $\dfrac{\text{change in speed (v – u)}}{\text{time taken (t)}}$ $a = \dfrac{(v - u)}{t}$

Acceleration is measured in metres per second per second (m/s/s), final speed (v) in metres per second (m/s), starting speed (u) in metres per second (m/s) and time in seconds (s).

For example: A car travels from 14 m/s to 30 m/s in 5 seconds. Calculate its acceleration.

$a = \dfrac{(v - u)}{t} = \dfrac{30 - 14}{5} = 3.2 \text{ m/s/s}$

● Motion can be shown graphically in distance–time and speed–time graphs.

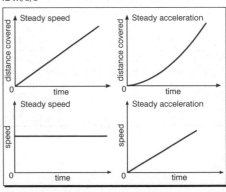

● The acceleration of an object depends on its mass and the force that is applied to it. The formula is:

force (F) = mass (m) × acceleration (a)   F = m a

Force is measured in newtons (N), mass in kilograms (kg) and acceleration in metres per second per second (m/s/s).

For example: A dragster has a resultant driving force of 5000 N and a mass of 400 kg. Calculate the dragster's initial acceleration.

$$a = \frac{F}{m} = \frac{5000}{400} = 12.5 \text{ m/s/s}$$

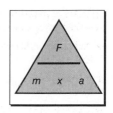

Stopping Distances

● When a car driver stops a car the **stopping distance** is made up of a **thinking distance** (the distance travelled in the time it takes the driver to react) and the **braking distance** (the distance travelled in the time it takes from beginning to brake until the car stops).

stopping distance = thinking distance + braking distance

● Thinking distances can vary from person to person and braking distances can vary from car to car.

Factors affecting thinking distance	Factors affecting braking distance
Speed	Speed
Tiredness	Tyre condition
Alcohol	Brake condition
Medication, drugs	Road condition
Level of concentration/ distraction	Mass of the car

FORCES AND MOTION (4)

Pressure

● **Pressure** is defined as the force per unit area:

$$\text{pressure (P)} = \frac{\text{force (F)}}{\text{area (A)}} \qquad P = \frac{F}{A}$$

Pressure is measured in newtons per metre squared (N/m^2), force in newtons (N) and area in metres squared (m^2).

For example: A statue weighs 600 N and has a base area of $0.25\,m^2$. Calculate the pressure the statue exerts on the ground.

$$P = \frac{F}{A} = \frac{600}{0.25} = 2400\,N/m^2$$

● Gases squash when pressure is applied. There are large spaces between the particles in a gas. This property is used in car air bags. The air bag allows a driver's head to decelerate slowly. This reduces the force exerted on the driver's neck and reduces the risk of injury. The relationship between the volume of a gas and its pressure at constant temperature is given by **Boyle's law**:

$$\text{pressure (P)} = \frac{\text{constant } (k)}{\text{volume (V)}}$$

$$PV = k \text{ or } P_1V_1 = P_2V_2$$

Pressure is measured in newtons per metre squared (N/m^2) and volume in decimetres cubed (dm^3) or litres.

For example: An air bag has a volume of $15\,dm^3$ under a pressure of $100\,000\,N/m^2$. The pressure is suddenly increased to $250\,000\,N/m^2$. What will be the new volume of the air bag?

$$V_2 = \frac{P_1V_1}{P_2} = \frac{100\,000}{250\,000} \times 15 = 6\,dm^3$$

● Liquids, unlike gases, cannot be compressed because their particles are too close together. This means that pressure can be transmitted through a liquid. This is the basis of **hydraulics**.

Forces and Motion (1–4)

1 A space craft weighs 5500 N on Earth but only 900 N on the Moon. Why are the weights different? (1)

2 Friction acts in a number of different places on a bicycle. Name one place where the friction is an advantage to the cyclist and one place where it is a disadvantage to the cyclist. (2)

3 A parachutist jumps out of a plane and after 5 seconds pulls open the parachute. The diagram shows the parachutist just after she has opened the parachute. Carefully describe her motion. (2)

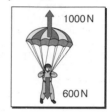

4 The stopping distance of a car depends on the thinking distance and the braking distance. List two factors that affect:

(a) the thinking distance. (1)

(b) the braking distance. (1)

5 A spring is stretched 10 cm by applying a force of 50 N. How much would the spring be stretched by a force of 80 N, assuming Hooke's Law applies? (2)

6 A skier accelerates from rest to 20 m/s in 8 seconds. Calculate the skier's overall acceleration. (3)

7 Look at the speed–time graph for a car. Describe the motion of the car at:

(a) Point A. (1)

(b) Point B. (1)

(c) Point C. (1)

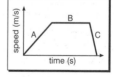

8 What force would be required to give an object of mass 3 kg acceleration of 10 m/s/s? (2)

9 The heel of a shoe exerts a pressure of 5000 N/m^2 on the ground. If the area of the heel is 0.020 m^2, calculate the force being applied. (3)

1 The force of gravity on the Moon is much less than that on the Earth. (1) The mass of the Moon is much less than that of the Earth; the gravity is about 1/6 of that on Earth.

2 Advantage: any one from friction between brake pad and wheel/ rider and seat/ rider and pedal/ wheel and road. (1) Disadvantage: any one from friction in wheel bearings/ in the chain/ in the pedals. (1)

3 She is moving downwards (1) but is decelerating (1). On opening the parachute the force due to air resistance will be greater than the force of gravity, but the parachutist is still falling to the ground.

4 (a) Any two from: speed/tiredness/ alcohol/ other drugs/ level of concentration. (1)
   (b) Any two from: speed/ tyre condition/ brake condition/ road condition/ mass. (1) Some exam questions ask about thinking time. Speed will not affect thinking time.

5 $F = k x$; $50 = k 10$ or $k = 5$. (1)
   $F = k x$; $80 = 5 x$ or $x = 16$ cm. (1)

6 Acceleration = $\dfrac{\text{change in speed}}{\text{time taken}}$ (1) = $\dfrac{20 - 0}{8}$ (1) = 2.5 m/s/s (1).
   The units of acceleration are rather unusual and are often given incorrectly by candidates.

7 (a) The car is accelerating. (1)
   (b) The car is travelling at constant speed. (1)
   (c) The car is decelerating. (1) You must be careful not to confuse speed–time graphs with distance–time graphs. A horizontal region on a distance–time graph indicates that the object is stationary.

8 $F = m a = 3 \times 10$ (1) = 30 N. (1) When using this formula remember that the mass is always given in kg.

9 $P = \dfrac{F}{A}$ or $F = P \times A$ (1) = 5000 × 0.02 (1) = 100 N. (1)

In all calculations it is important to write down the equation you are going to use and to show all your working. Also, don't forget the units!

TOTAL

Sources of Energy

- The **fossil fuels** of coal, oil and natural gas are still the most common sources of energy used in power stations. These fuels are **non-renewable**. They were formed over a period of millions of years and once supplies have been used up they cannot be replaced. Current stocks are likely to be used up in the next 300 years or so. Consequently it is important to develop other sources of energy: sources that are **renewable**.

Source of energy	Features
Solar power	The Sun's energy is transferred into electrical energy or is used to heat water.
Wind power	The wind is used to turn windmill-like turbines that generate electricity.
Wave and tidal power	Large floats move up and down with the waves and this movement is used to generate electricity.
Hydroelectric power	Water from dams is used to turn turbines and then generators, so producing electricity.
Biomass power	Plants use energy from the Sun during photosynthesis. The plant material can then be burnt, converted into fuels such as alcohol or used in biodigesters to make methane.
Geothermal power	In certain parts of the world water from hot springs can be used directly for heating.

- Unfortunately, the use of fossil fuels creates significant air pollution problems. However, their big advantage is that they have a very high energy content. The **alternative energy** sources listed in the table are generally much 'cleaner' but the efficiency of energy transfer is much less than that with fossil fuels. For example, several thousand windmill-turbines are needed to match the electricity output from a modern fossil fuel power station.

Transferring Energy

Thermal energy can be transferred in four main ways: conduction, convection, radiation and evaporation.

● **Conduction** is where thermal energy is transferred through a solid. If one end of a **conductor** is heated the atoms start to vibrate more vigorously. As the

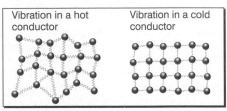

Vibration in a hot conductor

Vibration in a cold conductor

atoms are linked together by chemical bonds, the increased vibration can be passed on to other atoms. Air is a good **insulator**: because there are no bonds between the particles energy can only be transferred when the particles collide with each other.

● **Convection** occurs in fluids (liquids and gases). Hot fluids rise and cold fluids sink, creating **convection currents**. When part of a fluid is heated the particles move further apart, resulting in this part of the fluid becoming less dense than the unheated part. The heated part of the fluid then

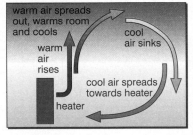

warm air spreads out, warms room and cools

cool air sinks

warm air rises

cool air spreads towards heater

heater

rises and the thermal energy is transferred. If a fluid's movement is restricted, energy cannot be transferred. That is why many insulators (e.g. ceiling tiles) have trapped air pockets.

● **Radiation**, unlike conduction and convection, does not need particles at all. Radiation can travel through a vacuum. The amount of energy radiated by an object depends on its temperature and its surface. Dull black surfaces are good absorbers and emitters of radiation; bright and shiny surfaces are poor absorbers and emitters of radiation.

ENERGY (3)

● **Evaporation** is when energetic molecules break away from the surface of a liquid. This reduces the average energy of the remaining molecules and so the liquid cools down.

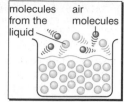

In the home a number of methods are used to reduce wasteful energy transfers. These include: cavity wall insulation, loft insulation, double glazing, draught excluders and carpets. Many of these methods seek to trap air and so reduce energy transfer by convection. When insulating a house you have to balance the cost of the insulation against the potential saving in energy costs. The **pay-back time** is the time it takes for the savings to repay the installation costs.

Work, Power and Energy

● **Work** is done when a force moves. It is calculated using the formula:
work done (W) = force (F) × distance moved (s) $W = F\,s$
Work done is measured in joules (J), force in newtons (N) and distance in metres (m).
Note: the distance must be measured in the direction of the force.
For example: A student with a weight of 500 N runs up a spiral staircase until she is 6 m above the ground. Calculate the work she does.
$W = F\,s = 500 \times 6 = 3000\,J$

● **Power** is the rate of doing work or the rate of transferring energy. Mechanical power (as opposed to electrical power, see *Electricity*) can be calculated using the formula:

power (P) = $\dfrac{\text{work done (energy transfer) (W)}}{\text{time taken (t)}}$ $P = \dfrac{W}{t}$

Power is measured in watts (W), work in joules (J) and time in seconds (s).
For example: A cyclist does 1500 J of work in 20 seconds.

Calculate his power. $P = \dfrac{W}{t} = \dfrac{1500}{20} = 75\,W$

ENERGY (4)

● Stored energy is known as **potential energy** (**P.E.**). If an object is lifted above the ground it will have **gravitational potential energy**, which will be transferred into **kinetic energy** (**K.E.**) if the object is allowed to fall to the ground. Potential energy depends on the weight of an object and its height above the ground. Kinetic energy depends on an object's mass and its velocity.

$$P.E. = m\,g\,h \qquad\qquad K.E. = \tfrac{1}{2}mv^2$$

P.E. and K.E are measured in joules (J), mass (m) in kilograms (kg), velocity (v) in metres per second (m/s), height (h) in metres and g is the gravitational field strength (approximately 10 N/kg).

For example: A skier of mass 56 kg is at the top of a ski run 2000 m above the finish line. Calculate her potential energy at the top of the ski run. Near to the bottom of the run she is travelling at a speed of 25 m/s. Calculate her kinetic energy. (The gravitational field strength is 10 N/kg.)

$$P.E. = m\,g\,h = 56 \times 10 \times 2000 = 1\,120\,000\,J \text{ or } 1120\,kJ$$

$$K.E. = \tfrac{1}{2}mv^2 = \tfrac{1}{2} \times 56 \times (25)^2 = 17\,500\,J \text{ or } 17.5\,kJ$$

At the top of the stone's flight, practically all of the stone's initial kinetic energy will have been transferred into potential energy. A small amount of energy will have been lost due to friction between the air and the stone.

● In most energy transfers some energy will end up as 'useless' heat. The **efficiency** of the transfer can be worked out using the equation:

$$\text{efficiency} = \frac{\text{useful energy output}}{\text{energy input}} \times 100\%$$

Energy (1–4)

1 Fossil fuels are 'non-renewable' sources of energy. What does this term mean? (1)

2 Name four renewable sources of energy. (1)

3 By what process is energy transferred in fluids such as liquids and gases? (1)

4 What type of surface will be a good absorber of radiation? (1)

5 What do you understand by the term 'pay-back time' in connection with the installation of an 'energy saving' device? (1)

6 Ceiling tiles can reduce energy transfer by conduction and convection. Explain how they do this. (2)

7 A pendulum transfers potential energy into kinetic energy and then kinetic energy back into potential energy.

   (a) At the top of a swing of a pendulum does the 'bob' have potential energy or kinetic energy or both? (1)

   (b) Explain, using ideas about energy transfer, why a pendulum eventually stops swinging. (2)

8 Explain how energy is transferred from the hot plate of an electric cooker to a potato cooking in water in a metal pan placed on the hot plate. (2)

9 Explain why evaporation causes cooling. (2)

10 A student with a weight of 600 N climbs 7 m up a vertical ladder in 10 seconds. Calculate:

   (a) the work done by the student. (2)

   (b) the power generated by the student during the climb. (2)

11 An archer fires an arrow of mass 100 g at a target. It hits the target at a speed of 9 m/s. Calculate the kinetic energy of the arrow as it hits the target. (2)

1 Fuels that take many millions of years to form and so cannot be replaced. (1) Do not say that they 'cannot be used again'. No fuel can be used again once it has been burnt!

2 Four from: solar; wind; tidal/wave; hydroelectric; biomass; geothermal; wood (certain types). (1)

3 Convection. (1)

4 Dull and black. (1) However, dull and black objects are also the best emitters of radiation.

5 This is the time it takes for savings on energy costs to compensate for the cost of installation. (1)
Double glazing tends to have a longer pay-back time but it also cuts down noise and prevents condensation on windows.

6 Conduction – they are made out of insulating material. (1)
Convection – they contain air which is trapped in the material. (1)

7 **(a)** Potential energy. (1)
When stationary, an object has zero kinetic energy.
(b) As potential energy is transferred into kinetic energy, and vice versa, energy is 'wasted' or 'lost' (1) due to friction between the pendulum holder and the string/bob and the air (1).

8 Any two from: energy is transferred from the hot plate to the pan by conduction (1); energy is transferred through the base of the pan by conduction (1); energy is transferred from the pan to the potato by convection through the water (1).

9 During evaporation the most energetic particles escape from the surface of the liquid into the air. (1) This reduces the average energy of the particles remaining in the liquid, resulting in a fall in temperature. (1)

10 **(a)** $W = F\,s$ (1) $= 600 \times 7 = 4200\,J$ (1).
If the ladder had not been vertical, the height used in the calculation would still have been the vertical height moved against the force of gravity (the student's weight).
(b) $P = \dfrac{W}{t}$ (1) $= \dfrac{4200}{10} = 420\,W$ (1).

11 K.E. $= \frac{1}{2}mv^2$ (1) $= \frac{1}{2} \times 0.1 \times 9^2 = 4.05\,J$ (1).
Remember the mass must be in kg.

TOTAL

The Properties of Waves

● There are two types of waves: **longitudinal** and **transverse**. Sound is an example of a longitudinal wave and travels as a result of particles vibrating in the same direction as the wave is travelling. There are regions of **rarefaction** and **compression**. Light and other electromagnetic waves are examples of transverse waves. In a transverse wave the vibrations are at right angles to the direction of the wave.

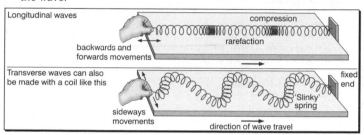

● The speed travelled by a wave depends on the substance or **medium** it is travelling through. All waves have the following features:
- they have a repeating shape
- they have a **frequency**, **wavelength** and **amplitude**
- they carry energy without moving the material along.

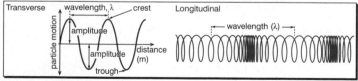

Characteristic of a wave	Definition
Wavelength	The length of the repeating pattern
Frequency	The number of repeating patterns which pass a point each second
Amplitude	The maximum displacement of the medium's vibration

WAVES (2)

- The speed of a wave in a given medium is constant. The relationship between speed, wavelength and frequency is given by the formula:

wave speed (v) = frequency (f) × wavelength (λ) $v = f\lambda$

Wave speed is measured in metres per second (m/s), frequency in hertz (Hz) and wavelength in metres (m).

For example: A loudspeaker makes sound waves with a frequency of 600 Hz. The waves have a wavelength of 0.55 m. Calculate the speed of the sound waves.

$v = f\lambda = 600 \times 0.55 = 330$ m/s

- When waves hit a barrier they are **reflected** and the angle of incidence equals the angle of reflection.

- When a wave slows down the **wavefronts** crowd together and the wavelength gets smaller. If a wave enters a different medium at an angle the wavefronts change direction. This is known as **refraction**.

- Wavefronts change shape when they go through a gap. This process is known as **diffraction**. Diffraction is most noticeable when the size of the gap equals the wavelength of the waves.

- Earthquakes make **seismic waves**. These are made up of longitudinal primary waves (**P-waves**) and transverse secondary waves (**S-waves**). P-waves can travel through liquid and solid whereas S-waves can only travel through solid rock. Evidence from earthquakes suggests that part of the Earth's core is made of liquid.

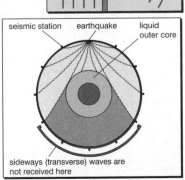

seismic station   earthquake   liquid outer core

sideways (transverse) waves are not received here

The Electromagnetic Spectrum

● The **electromagnetic spectrum** is a family of different kinds of transverse waves. The waves all travel at the same speed in a vacuum (the speed of light, 300 000 000 m/s) but have different wavelengths and frequencies.

Type of wave	gamma rays	X rays	ultraviolet	visible	infrared	microwaves	TV and radio waves
frequency	high						low
wavelength	low						high
use	killing cancer cells	to look at bones	sun tan beds	photography	TV remote controls	cooking	transmission of TV and radio

● The **visible spectrum**, part of the electromagnetic spectrum, can be produced by passing a beam of white light through a prism. As the different colours in the visible spectrum have different wavelengths and frequencies, they are refracted by different amounts as they pass through the prism.

Light

● Light is **reflected** at shiny polished surfaces such as mirrors. When reflection occurs the angle of incidence equals the angle of reflection. The angles of incidence and reflection are measured to an imaginary line drawn at right angles to the mirror surface, known as the **normal**.

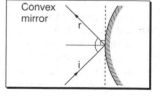

● Light is **refracted** when it travels from air into glass and vice versa. Lenses are commonly used to refract light.

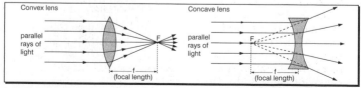

Type of mirror	Type of image	Uses
Plane	Same size as object	Household 'dressing' mirror, periscope
Concave	Magnified (if object close to mirror)	Make up/ shaving mirrors
Convex	Smaller than the object	Car driving mirror, shop security mirror

● When rays of light pass from a dense medium into a less dense medium they are refracted away from the normal. When the angle of incidence exceeds a certain value the ray is reflected internally rather than being refracted. This process is known as **total internal reflection** and the angle of incidence at which refraction stops is called the **critical angle**. This explains how fibre optic cables work.

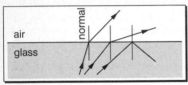

Sound

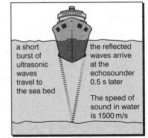

a short burst of ultrasonic waves travel to the sea bed

the reflected waves arrive at the echosounder 0.5 s later

The speed of sound in water is 1500 m/s

● Sound is a longitudinal wave that travels at a speed of 340 m/s in the air. Sound is caused by vibrations. The frequency of a sound wave is called its **pitch**; the amplitude is the **loudness** of the sound and is measured in **decibels**. Sound with a frequency above the range the human ear can detect is known as **ultrasound**.

● Ultrasound is used in echo sounding or **SONAR**.

For example: A ship sends out an ultrasound wave and receives an echo in 0.5 s. What is the depth of the water?

Distance = speed × time = $1500 \times 0.5 = 750$ m

Depth = $\frac{1}{2}$ distance = 375 m.

Waves (1–4)

1 Are electromagnetic waves transverse or longitudinal? (1)

2 The parts of the electromagnetic spectrum are shown below:

Radio	A	Visible	B	X-rays	γ-rays

   (a) What is the name of part A? (1)

   (b) What is the name of part B? (1)

   (c) Which waves have the higher frequency, radio or X-rays? (1)

   (d) Which waves have the higher wavelength, radio or γ-rays? (1)

   (e) Give one use of γ-rays. (1)

   (f) What do all the parts of the electromagnetic spectrum have in common? (1)

3 Look at the drawing of a transverse wave.

   (a) Which letter shows the trough of the wave? (1)

   (b) Which letter shows the amplitude of the wave? (1)

   (c) The wavelength is the distance between which two letters? (1)

4 Give one use of each of the following:

   (a) a convex mirror. (1)

   (b) a concave mirror. (1)

5 Look at the diagram, which shows a ray of light passing from medium A into medium B.

   (a) What name is given to the bending of light in this way? (1)

   (b) Which medium is more dense, A or B? Explain your answer. (1)

6 Explain how total internal reflection occurs. (3)

7 A destroyer detects a submarine below it using sonar. The destroyer sends out an ultrasound beam and receives an echo from the submarine in 0.8 s. Calculate the depth of the submarine (the speed of sound in water is 1500 m/s). (3)

1 Transverse. (1)
   Sound waves are the most common longitudinal waves.

2 (a) Infra-red. (1)
   (b) Ultra-violet. (1)
   (c) X-rays. (1) Higher frequency waves have more energy.
   (d) Radio waves. (1) As the speed of a wave, $v = f\lambda$, and v
       is constant, waves with low frequency have high
       wavelength.
   (e) Killing cancer cells. (1) γ-rays have very high frequency
       and energy and have great penetrating power (higher
       than X-rays). They have to be directed very carefully at
       the cancer cells as they can kill healthy cells as well.
       Other uses include sterilising equipment/food and in a
       gamma camera.
   (f) They all travel at the same speed. (1) This speed is the
       speed of light.

3 (a) C. (1) This is the opposite of the crest (A or E).
   (b) F. (1)
   (c) A and E. (1) The wavelength must include the whole
       repeating pattern.

4 (a) One from: car driving mirror/ shop security mirror. (1)
       This gives an image smaller than the object.
   (b) One from: make-up/ shaving mirror. (1) This gives a
       magnified image.

5 (a) Refraction. (1)
   (b) Medium B is the more dense. Light bends towards the
       normal when passing into a denser medium. (1)
       When light travels from a more dense to a less dense
       medium it bends away from the normal.

6 If light hits a surface at an angle greater or equal (1) to the
   critical angle (1), then it will be reflected rather than
   refracted (1).

7 Distance = speed × time. (1) $s = 1500 \times 0.8 = 1200\,\text{m}$. (1)
   The depth of the submarine = $\frac{1}{2}$ distance travelled by the
   wave = $\frac{1}{2} \times 1200 = 600\,\text{m}$. (1)
   In these types of question don't forget that when the echo is
   received the wave has travelled to the object and back again.

TOTAL

The Solar System

● The **solar system** is made up of the Sun and nine planets that orbit it. The planets are attracted to the Sun by the force of gravity and orbit it in oval or elliptical orbits.

Mercury Venus Earth Mars Jupiter Saturn Uranus Neptune Pluto

Increasing distance from the Sun

● **Asteroids** are fragments of rock that orbit the Sun between the four inner planets and five outer planets.

● **Comets** are made from frozen ice and rock and orbit the Sun in very elliptical orbits. Most of the time the comets cannot be seen because they are too far away from the Sun. When they get closer some of the solid turns into gas and forms a 'tail'.

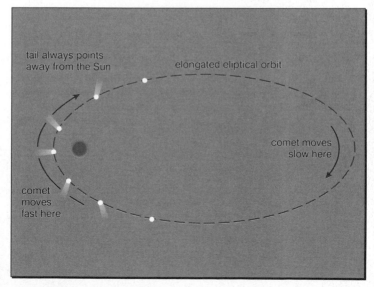

tail always points away from the Sun

elongated eliptical orbit

comet moves slow here

comet moves fast here

- Natural and artificial satellites orbit the Earth due to a **centripetal force**: a force that constantly pulls towards the centre. This force is gravity. Artificial satellites can orbit the Earth without needing motors or rockets. Some satellites are in **geo-stationary** orbits. They appear to be stationary but are orbiting the Earth at exactly the same speed as the Earth is rotating, so they remain above the same part of the Earth.

- The Sun is part of a group of stars called a **galaxy**. Our Sun is part of the **Milky Way** galaxy, which is made up of billions of stars. The Milky Way is just one of billions of galaxies which make up the **Universe**.

The Life-cycle of a Star

- A star starts as a cloud of hydrogen gas. Gravity pulls the hydrogen atoms closer together, creating high temperatures that allow **nuclear fusion** to take place. The core sends out light and the cool, outer layers are blown away, sometimes forming planets around the new star.

- The star is held together as the outward force exerted by the hot gases is exactly balanced by the inward force of gravity. Some stars appear red–orange in colour but hotter stars appear blue–white (**blue stars**).

- As the hydrogen in the star's core starts to get used up, the core starts to cool and gravity causes the star to start to collapse. New nuclear fusion reactions start and the star expands to form a **red giant** or **red supergiant**. Eventually all the nuclear fusion reactions will be complete. In a medium-sized star like the Sun the core collapses and the red giant forms a **white dwarf star**; in a large star the red supergiant forms a **supernova**.

- A supernova is brighter than a whole galaxy of stars. The very dense core becomes a **neutron star** or, if it is very small and dense, a **black hole**.

- A cloud of gas where stars are being formed is known as a **nebula**. Many nebulas are also formed by stars exploding.

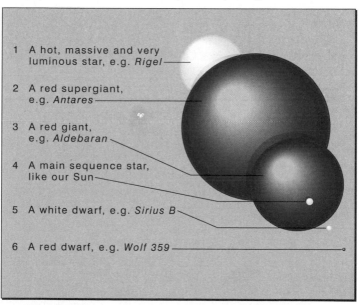

1 A hot, massive and very luminous star, e.g. *Rigel*

2 A red supergiant, e.g. *Antares*

3 A red giant, e.g. *Aldebaran*

4 A main sequence star, like our Sun

5 A white dwarf, e.g. *Sirius B*

6 A red dwarf, e.g. *Wolf 359*

The Origin of the Universe

● Evidence for the way the Universe formed has been obtained by analysing the frequency of light obtained from distant galaxies.

● The frequency of light obtained from a distant star is affected by the movement of the star. If the star is moving away from the Earth, light waves reach the Earth with a lower frequency than those emitted from the star. This shift to lower frequency is know as a **red shift**. Light from all distant galaxies shows a red shift, proving that all these galaxies are moving away from the Earth. In other words the Universe is expanding.

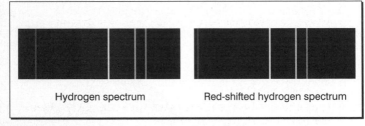

Hydrogen spectrum   Red-shifted hydrogen spectrum

● One explanation for the fact that the galaxies are moving apart from each other is known as the **Big Bang** theory. The Big Bang theory suggests that everything in the Universe originated from the same point and then exploded apart. Since this explosion, 11–18 billion years ago, the Universe has been expanding and cooling.

● Astronomers suggest that there are three main possibilities for the future of the Universe:
  ■ the Universe will keep on expanding
  ■ the expansion will slow down and stop
  ■ the Universe will start to contract.

These all depend on what the balance is between 'expansion forces' and gravitational forces.

The Earth and Beyond (1–4)

1 What is the name of the planet that is:

(a) closest to the Sun? (1)

(b) furthest away from the Sun? (1)

(c) closest to the Earth? (1)

2 Where would you find the asteroid belt? (1)

3 Arrange the following in order of size, starting with the smallest first:

solar system   star   universe   galaxy (1)

4 What is a geo-stationary orbit? (1)

5 What type of reaction occurs in the centre of a star? (1)

6 What is a:

(a) nebula? (1)

(b) supernova? (1)

(c) white dwarf? (1)

7 Explain what is meant by a red shift. What information does this give about distant galaxies? (2)

8 Explain what is meant by the Big Bang theory. (2)

9 List three possibilities for the future of the Universe. For each possibility explain the forces that would be acting on the Universe. (2)

10 Describe the processes involved in the 'death' of a medium-sized star such as the Sun. (4)

1 **(a)** Mercury. (1)
 (b) Pluto. (1)
 (c) Venus. (1)
 The Earth is just over 40 million km from Venus and is about 80 million km from Mars.
2 The asteroid belt is found between the four inner planets and the five outer planets, i.e. between Mars and Jupiter. (1)
3 star, solar system, galaxy, universe. (1)

4 An object orbiting at exactly the same speed as the Earth is rotating. (1) They are very useful for global communication. Signals can be 'bounced off' them.
5 Nuclear fusion. (1)
 This involves the joining together of two hydrogen nuclei to form a helium nucleus. In nuclear fission nuclei are broken apart.
6 **(a)** A nebula is a cloud of gas formed when stars explode or where new stars are being formed. (1)
 (b) A supernova is formed when a large star explodes. (1)
 (c) A white dwarf is formed when a medium-sized star collapses. (1) This collapse occurs when the nuclear reactions stop.
7 A red shift is when the lines in a spectrum move towards the red end of the spectrum, i.e. to a lower frequency. (1) This shows that distant galaxies are moving away from the Earth. (1)
8 The Big Bang theory suggests that everything in the Universe started the same point (1) and then exploded apart sending galaxies moving outwards from that point (1).
9 The Universe: continues to expand; the expansion slows down and stops; starts to contract. (1) In the expanding Universe model the expansion force is greater than the force of gravity; for the expansion to stop the expansion forces must equal the gravity forces; for the contraction to occur the gravity forces must be greater than the expansion forces. (1)
10 The hydrogen in the star's core gets used up. (1) The star starts to collapse before other nuclear reactions cause it to expand rapidly (1) to form a red giant. (1) All nuclear reactions eventually finish and the star collapses to form a white dwarf star. (1)

TOTAL

RADIOACTIVITY (1)

Types of Radiation

- An atom is made up of a nucleus (containing protons and neutrons) surrounded by 'shells' of electrons. The protons in the nucleus will repel each other, so neutrons can be thought of as particles which separate the protons and reduce this repulsion. Some atoms are unstable and 'decay' or break down into more stable nuclei. This process is known as **nuclear fission**.

- When atoms decay they give out particles, radiation and energy: they are said to be **radioactive**. There are three main types of radiation: **alpha** (α), **beta** (β) and **gamma** (γ). The radiation they produce is known as **ionising radiation**.

- α particles are positively charged, made up of two protons and two neutrons (a helium nucleus). β particles are negatively charged, identical to an electron. γ rays are uncharged; they are a type of electromagnetic radiation.

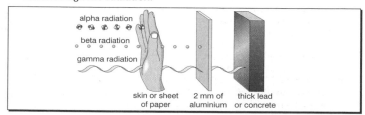

alpha radiation

beta radiation

gamma radiation

skin or sheet of paper | 2 mm of aluminium | thick lead or concrete

- Radioactivity is measured using a **Geiger-Muller tube** linked to a counter.

G–M tube

- Even when a radioactive source is not being used there will always be a certain amount of ionising radiation present. This is known as **background radiation**. It is caused by:
  - radioactivity in soils, rocks and other materials
  - radioactive gases (e.g. radon)
  - cosmic rays from the Sun.

Half-Life

- A sample of a radioactive material decays over time. The **activity** of the sample (the number of ionising particles it emits each second) will decrease as the unstable atoms present change into stable atoms. The time taken for half the radioactive atoms to decay is known as the **half-life**. In radioactive decay the half-life is a constant value for a given radioactive element or **isotope**.

- The graph shows a typical decay curve. The activity decreases by half every two days (the number of radioactive atoms decreases by half every two days). The half-life is therefore two days.

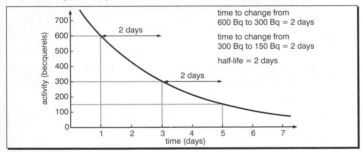

For example: A radioactive isotope is monitored using a Geiger-Muller tube and counter. Initially it has an activity of 800 counts per minute. After four hours the activity has dropped to 50 counts per minute. What is the half-life of the isotope?

Half-lives	Count rate (counts/min)	Each halving of the activity is one half-life.
0	800	
1	400	
2	200	
3	100	
4	50	4 hours is the same as 4 half-lives.
Half-life = 1 hour.		

- Different isotopes have different half-lives. Uranium has a half-life of 4500 million years. Half of the uranium atoms will have turned into lead atoms after 4500 million years. Carbon-14, which is used in **radioactive carbon dating**, has a half-life of 5700 years.

Nuclear Equations

- Nuclear reactions can be represented by equations. Each nucleus is represented by its chemical symbol, and the atomic number and mass number are written in front of the symbol:

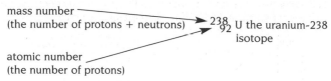

mass number
(the number of protons + neutrons) $\longrightarrow$ $^{238}_{92}$ U the uranium-238 isotope

atomic number
(the number of protons)

- In a nuclear equation the mass numbers and atomic numbers must balance on both sides of the equation:

For example:

α decay $\qquad$ $^{226}_{88}$ Ra $\rightarrow$ $^{222}_{86}$ Rn + $^{4}_{2}\alpha$

β decay $\qquad$ $^{218}_{84}$ Po $\rightarrow$ $^{218}_{85}$ At + $^{0}_{-1}\beta$

Nuclear Fission

- The energy produced in a nuclear power station comes from the decay of unstable nuclei in a process known as **nuclear fission**. Atoms of uranium-235 are bombarded with neutrons and split into two smaller atoms. The fission is accompanied by the release of a large amount of energy. One kilogram of uranium-235 produces the same amount of energy as 2 000 000 kg of coal. The products of nuclear fission are highly radioactive and need to be disposed of very carefully.

Nuclear Fusion

● Energy is also released when small atoms join or fuse together in a process known as **nuclear fusion**. This is the process that occurs in the Sun. Scientists are trying to develop nuclear fusion processes that can be used to generate electricity.

The Uses of Radioactivity

Radioactive isotopes have a number of important uses.

Use	Key features
Sterilising	Gamma rays kill bacteria. They can be used to sterilise medical equipment and preserve food.
Smoke detectors	Smoke detectors contain an alpha radiation source. Smoke reduces the number of available alpha particles so setting off the alarm.
Thickness measurement	Beta particles are used to monitor the thickness of paper or metal.
Checking welds	A 'gamma camera' is used to identify faults in welds.
Medical	Gamma rays destroy cancer cells.
Tracers	Tracers can be used to detect blockages in vital human organs, to monitor nutrient flow in plants and to detect leaks in pipes.

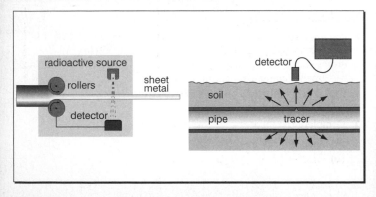

Radioactivity (1–4)

1 Answer the questions by selecting answers from the list below.

Which form of ionising radiation:

| α particles |
| β particles |
| γ rays |

(a) is made up of electrons? (1)

(b) is made up of helium nuclei? (1)

(c) will not pass through skin? (1)

(d) is used to sterilise medical equipment? (1)

(e) is used for controlling the thickness of thin sheet metal? (1)

(f) is emitted by the radioactive source in a smoke alarm? (1)

2 What is the name of the device that can be used to measure the activity of a radioactive source? (1)

3 What is the process that generates energy in the Sun? (1)

4 What is the name of the process that is used in nuclear power stations to generate electricity? (1)

5 There is always a certain amount of background radiation present. Give two sources of this background radiation. (2)

6 (a) Use the radioactive decay curve to work out the half-life of the radioactive source. (2)

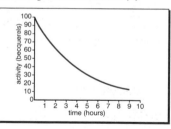

(b) Another radioactive source has a half life of 6 hours. If its activity is 2000 counts/min at 12 noon what will be its count rate 24 hours later? (3)

7 Complete the following nuclear equations:

(a) $^{232}_{90}\text{Th} \rightarrow {}^{x}_{y}\text{Ra} + {}^{4}_{2}\alpha$ (2)

(b) $^{14}_{y}\text{C} \rightarrow {}^{x}_{7}\text{N} + {}^{0}_{-1}\beta$ (2)

1 (a) β particles. (1)
   (b) α particles. (1)
   (c) α particles. (1) Alpha particles are the least penetrating of the three.
   (d) γ rays. (1) These are very high energy electromagnetic waves.
   (e) β particles. (1) Beta particles are used as their passage through paper or thin metal is likely to change significantly even with a small change in thickness. Gamma rays would be much less sensitive and also more hazardous to use. (1)
   (f) α particles. (1)

2 Geiger-Muller tube. (1)
   'Geiger counter' is in more everyday usage but try and remember the full title of the equipment.

3 Nuclear fusion. (1)
   In this process small nuclei join together to release large quantities of energy.

4 Nuclear fission. (1)
   In this process large nuclei break into smaller nuclei usually as a result of being bombarded with neutrons.

5 Any two from: radioactivity in soil/ rocks/ materials such as concrete; radioactive gases such as radon; cosmic rays from the Sun. (2)

6 (a) Half-life = 3 hours. (2)
       The activity falls from 100 to 50 in 3 hours and from 50 to 25 in a further 3 hours, etc.
   (b) 24 hours = 24/6 = 4 half-lives. (1) Therefore the initial count rate will halve 4 times. (1) $2000 \rightarrow 1000 \rightarrow 500 \rightarrow 250 \rightarrow 125$ counts/ min. (1)

7 (a) $x = 228$. (1) The mass number totals on both sides of the equation must be the same, i.e. $232 = 228 + 4$.
       $y = 88$. (1) The atomic number totals on both sides of the equation must be the same, i.e. $90 = 88 + 2$.
   (b) $x = 14$. (1) $14 = 14 + 0$. $y = 6$. (1) $6 = 7 - 1$.

TOTAL

Topic	Check yourself	Points out of 20
Life Processes and Cells	1	20
Human Body Systems	2	20
Body Maintenance	3	20
Plants	4	20
Ecology and the Environment	5	20
Genetics and Evolution	6	20
Formulae and Equations	7	20
Structure and Bonding	8	20
Fuels and Energy	9	20
Rocks and Metals	10	20
Chemical Reactions	11	20
The Periodic Table	12	20
Chemical Calculations	13	20
Electricity	14	20
Electromagnetism	15	20
Forces and Motion	16	20
Energy	17	20
Waves	18	20
The Earth and Beyond	19	20
Radioactivity	20	20

SCORE CHART (2)

Mark your points for each *Check yourself* on the grid and then read across for your grade.

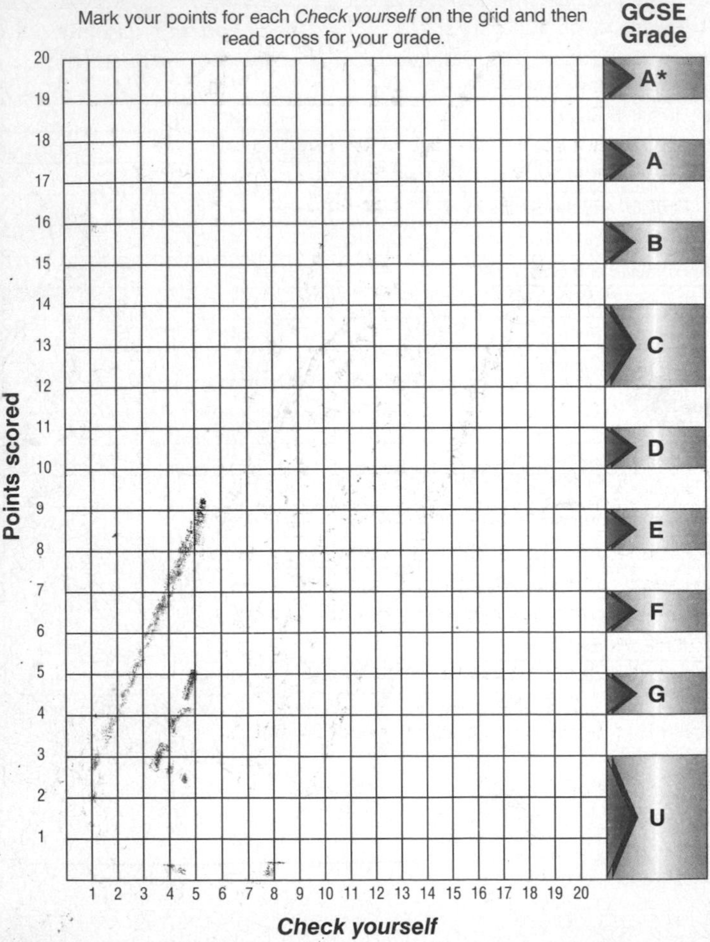
GCSE
Grade
A*
A
B
C
D
E
F
G
U
Points scored
20
19
18
17
16
15
14
13
12
11
10
9
8
7
6
5
4
3
2
1
1 2 3 4 5 6 7 8 9 10 11 12 13 14 15 16 17 18 19 20
Check yourself